Springer Series in Wood Science
Editor: T. E. Timell

Martin H. Zimmermann
Xylem Structure and the Ascent of Sap (1983)

John F. Siau
Transport Processes in Wood (1984)

John F. Siau

Transport Processes in Wood

With 123 Figures

Springer-Verlag
Berlin Heidelberg New York Tokyo 1984

JOHN FINN SIAU
Professor of Wood Products Engineering
Department of Wood Products Engineering
SUNY College of Environmental Science and Forestry
Syracuse, New York 13210, U.S.A.

Series Editor:

T. E. TIMELL
State University of New York
College of Environmental
Science and Forestry
Syracuse, NY 13210
U.S.A.

Cover: Transverse section of *Pinus lambertiana* wood. Courtesy of Dr. Carl de Zeeuw, SUNY College of Environmental Science and Forestry, Syracuse, New York

---

ISBN 3-540-12574-4 Springer-Verlag Berlin Heidelberg New York Tokyo
ISBN 0-387-12574-4 Springer-Verlag New York Heidelberg Berlin Tokyo

Library of Congress Cataloging in Publication Data Siau, John F. Transport processes in wood. (Springer series in wood science; v. 2) Based on: Flow in wood, 1971. 1. Wood–Permeability. 2. Wood–Moisture. 3. Wood–Anatomy. I. Title. II. Series. TA419.S426 1983 674′.132 83-10457 ISBN 0-387-12574-0 (U.S.)

This work is subject to copyright. All rights are reserved, whether the whole or part of the material is concerned, specifically those of translation, reprinting, re-use of illustrations, broadcasting, reproduction by photocopying machine or similar means, and storage in data banks. Under § 54 of the German Copyright Law, where copies are made for other than private use, a fee is payable to "Verwertungsgesellschaft Wort", Munich

© by Springer-Verlag Berlin Heidelberg 1984
Printed in Germany.

The use of registered names, trademarks, etc. in this publication does not imply, even in the absence of a specific statement, that such names are exempt from the relevant protective laws and regulations and therefore free for general use.

Typesetting, printing and bookbinding: Brühlsche Universitätsdruckerei, Giessen
2131/3130-543210

Preface

This book has a similar subject content to the author's previous *Flow in Wood* but with substantial updating due to the abundance of research in the wood science field since 1971. Several different concepts have been introduced, particularly in regard to wood–moisture relationships. The role of water potential in the equilibria between wood and its humid and moist environments is considered. Two theories are introduced to explain the nonisothermal transport of bound water in the steady and unsteady states. As in the former text, the wood-structure relationship is emphasized.

The author is especially grateful to Dr. C. Skaar for his careful and critical review of much of the manuscript and for the productive discussions of many of the concepts. Dr. T. E. Timell, the series editor, rendered major assistance in the preparation of Chap. 2 and in his editing of the manuscript. The author wishes to thank Dr. W. A. Côté, Mr. A. C. Day, and Mr. J. J. McKeon for providing electron micrographs, Mr. G. A. Snyder for his photography of much of the art work, Dr. C. H. de Zeeuw for his advice in the field of wood anatomy, and Ms. Mary M. Siau for her careful rendition of the art work. Appreciation is extended to Miss Judy A. Barton and Mrs. Stephanie V. Micale for their work in typing and checking the manuscript. Mr. J. A. Meloling of Carrier Corporation provided the psychrometric charts, and the U.S.D.A. Forest Service furnished several items of art work. Finally, the author wishes to thank Dr. G. H. Kyanka, Dean W. P. Tully, and President E. E. Palmer of SUNY College of Environmental Science and Forestry for permitting the expenditure of time and the use of college facilities for writing this book.

Syracuse, N.Y. JOHN F. SIAU
November 1983

Contents

1 Basic Wood-Moisture Relationships

1.1 Introduction 1
1.2 Saturated Vapor Pressure 1
1.3 Relative Humidity 5
    1.3.1 Use of the Psychrometric Chart 7
    1.3.2 Measurement of Relative Humidity 7
    1.3.3 Control of Relative Humidity 18
1.4 Equilibrium Moisture Content and the Sorption Isotherm 19
1.5 The Effect of Changes in Pressure and Temperature on Relative Humidity 24
1.6 Specific Gravity and Density 25
1.7 Specific Gravity of the Cell Wall and Porosity of Wood . 25
1.8 Swelling and Shrinkage of the Cell Wall 30
1.9 Swelling and Shrinkage of Wood 31

2 Wood Structure and Chemical Composition

2.1 Introduction 35
2.2 The Cell Wall 36
2.3 Structure of Softwoods 40
2.4 Types of Pit Pairs 45
2.5 Softwood Pitting 49
2.6 Microscopic Studies of Flow in Softwoods 52
2.7 Structure of Hardwoods 53
2.8 Hardwood Pitting 58
2.9 Microscopic Studies of Flow in Hardwoods 61
2.10 Chemical Composition of Normal Wood 63
    2.10.1 Cellulose 63
    2.10.2 Hemicelluloses 65
        2.10.2.1 Introduction 65
        2.10.2.2 Softwood Hemicelluloses 66
        2.10.2.3 Hardwood Hemicelluloses 67
    2.10.3 Lignins 68
2.11 Chemical Composition of Reaction Wood 70
    2.11.1 Introduction 70
    2.11.2 Compression Wood 70
    2.11.3 Tension Wood 71
2.12 Topochemistry of Wood 72

3 Permeability

3.1 Introduction 73
3.2 Darcy's Law 73
3.3 Kinds of Flow 76
3.4 Specific Permeability 78
3.5 Poiseuille's Law of Viscous Flow 79
3.6 Turbulent Flow 82
3.7 Nonlinear Flow Due to Kinetic-Energy Losses at the Entrance of a Short Capillary 83
3.8 Knudsen Diffusion or Slip Flow 85
3.9 Corrections for Short Capillaries 88
3.10 Permeability Models Applicable to Wood 89
    3.10.1 Simple Parallel Capillary Model 89
    3.10.2 Petty Model for Conductances in Series . . . 89
    3.10.3 Comstock Model for Softwoods 91
    3.10.4 Characterization of Wood Structure from Permeability Measurements 93
3.11 Measurement of Liquid Permeability 95
3.12 Measurement of Gas Permeability 95
3.13 The Effect of Drying on Wood Permeability 97
3.14 Treatments to Increase Permeability 98
3.15 The Effect of Moisture Content on Permeability . . 100
3.16 The Influence of Specimen Length on Permeability . 100
3.17 Permeability of the Cell Wall 102
3.18 Zones of Widely Differing Permeabilities in Wood . 103
3.19 General Permeability Variation with Species 103

4 Capillary and Water Potential

4.1 Surface Tension 105
4.2 Capillary Tension and Pressure 105
4.3 Mercury Porosimetry 108
4.4 Influence of Capillary Forces on the Pressure Impregnation of Woods with Liquids 111
4.5 Collapse in Wood 111
4.6 Pit Aspiration 115
4.7 The Relationship Between Water Potential and Moisture Movement 118
4.8 Notes on Water Potential. Equilibrium Moisture Content, and Fiber Saturation Point of Wood 124

5 Thermal Conductivity

5.1 Fourier's Law 132
5.2 Empirical Equations for Thermal Conductivity . . . 134
5.3 Conductivity Model 135
5.4 Resistance and Resistivity; Conductance and Conductivity . 136
5.5 Derivation of Theoretical Transverse Conductivity Equation 137

5.6 Derivation of Theoretical Longitudinal Conductivity
     Equation 143
5.7 R and U Values; Convection and Radiation 146
5.8 Application to Electrical Resistivity Calculations 149
5.9 Application to Dielectric Constant Calculations 149

6 Steady-State Moisture Movement

6.1 Fick's First Law Under Isothermal Conditions 151
6.2 Bound-Water Diffusion Coefficient of Cell-Wall Substance 152
6.3 The Combined Effect of Moisture Content and Temperature
     on the Diffusion Coefficient of Cell-Wall Substance . . . 155
6.4 Water-Vapor Diffusion Coefficient of Air in the Lumens . 157
6.5 The Transverse Moisture Diffusion Model 160
6.6 The Importance of Pit Pairs in Water-Vapor Diffusion . 161
6.7 Longitudinal Moisture Diffusion Model 163
6.8 Nonisothermal Moisture Movement 165
6.9 Measurement of Diffusion Coefficients by Steady-State
     Method 171

7 Unsteady-State Transport

7.1 Derivation of Unsteady-State Equations for Heat and
     Moisture Flow 175
7.2 Derivation of Unsteady-State Equations for Gaseous Flow
     in Parallel-Sided Bodies 179
7.3 Graphical and Analytical Solutions of Diffusion-Differential Equations with Constant Coefficients 183
     7.3.1 Solutions of Equations for Parallel-Sided Bodies . 185
     7.3.2 Solutions of Equations for Cylinders 190
     7.3.3 Simultaneous Diffusion in Different Flow Directions 194
     7.3.4 Significance of Flow in Different Directions 195
     7.3.5 Special Considerations Relating to the Heating of
           Wood 203
7.4 Relative Values of Diffusion Coefficients 206
7.5 Retention 210
7.6 Unsteady-State Transport of Liquids 213
     7.6.1 Parallel-Sided Bodies, Permeability Assumed
           Constant with Length 213
     7.6.2 Parallel-Sided Bodies with Permeability Decreasing
           with Length (Bramhall Model) 216
     7.6.3 Cylindrical Specimens 218
     7.6.4 Square and Rectangular Specimens 219
7.7 Unsteady-State Transport of Moisture Under Nonisothermal Conditions 221
7.8 Heat Transfer Through Massive Walls 222

References 226

Symbols and Abbreviations 233

Subject Index 237

Chapter 1
Basic Wood–Moisture Relationships

1.1 Introduction

Moisture in wood exists in two basic forms: *bound water* within the cell wall and *free water* in liquid form in the voids of the wood. An equilibrium is attained between the bound-water content of the wood and the relative humidity of the surrounding air such that at a relative humidity of 0% the wood moisture content is essentially zero. This *equilibrium moisture content* gradually increases with the relative humidity until the cell wall becomes saturated when the relative humidity approaches 100%. Beyond this point, called the *fiber saturation point*, additional water is in the free form. An equilibrium also exists between wood and moist soil in contact with it where the equilibrium moisture content includes free water. Green wood contains free water which is removed by exposing it to warm, dry air. The presence of free water in wood drastically increases its decay susceptibility.

Nearly all the physical properties of wood are influenced significantly by its moisture content. As the bound-water content increases wood swells, mechanical strength decreases, thermal and electrical conductivities increase, and the rate of bound-water diffusion increases. These changes are usually gradual and continuous until the cell wall is saturated, beyond which there is little additional change in these properties. Because the moisture content of wood up to the fiber saturation point is of the greater importance in its effect on properties and transport processes, an understanding of the humidity of the air, the factors affecting it, and the means of its control provides a necessary background for the study of permeability, capillary behavior, thermal conductivity, and moisture diffusion.

1.2 Saturated Vapor Pressure

Assume that a reservoir of water is placed in an enclosure initially containing dry air or a vacuum, as illustrated in Fig. 1.1. The liquid molecules are in a state of constant vibration and have average kinetic energies directly proportional to their absolute temperatures. The molecules possess a distribution of energies representing a large range of values. A very small fraction of the molecules will have sufficient energy to escape from the surface and form vapor molecules in the space above the reservoir. This evaporation process will continue and will be counteracted by condensation of some of the molecules. Eventually a steady state is achieved when the rate of evaporation and condensation become equal. At this time the concentration of vapor molecules results in a partial vapor pressure known as the saturated vapor pressure (p_o). This pressure will be the same regardless of the amount of air present,

2   Basic Wood–Moisture Relationships

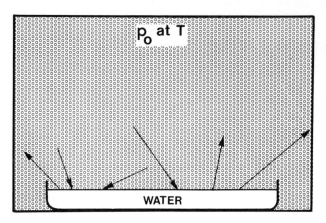

Fig. 1.1. After equilibrium has been established at a given temperature in a confined space, the partial vapor pressure will be equal to the saturated vapor pressure

provided the temperature remains constant. This is in accordance with Dalton's law of partial pressure, which states that each constituent of a mixture of perfect gases behaves as it it were present alone at the temperature of the mixture. Thus the pressure of the atmosphere (P_a) can be considered to be the sum of the partial pressures of dry air and water vapor.

$$P_a = p_0 + p_{air} \tag{1.1}$$

where P_a = atmospheric pressure, p_0 = saturated vapor pressure, p_{air} = partial vapor pressure of air.

The evaporation of water is a temperature-activated process and, as such, the saturated vapor pressure may be calculated with relatively good precision from an equation of the form

$$p_0 = 8.75 \times 10^7 \exp(-10{,}400/RT), \tag{1.2}$$

where p_0 = saturated vapor pressure, cm Hg, 10,400 cal/mol = escape energy, R = universal gas constant = 2 cal/mol K, T = Kelvin temperature.

It is clear from Eq. (1.2) that saturated vapor pressure is dependent on temperature only since the escape energy and R are constants, and it is therefore independent of the ambient air pressure. The escape energy is the molecular heat of vaporization and, dividing by the molecular weight of water (18 g/mol) the value of 578 cal/g is obtained, which is the heat of vaporization at approximately 40 °C. The term, RT, is proportional to the average thermal energy per mole which is therefore directly proportional to the absolute temperature. Thus temperature may be defined in terms of average molecular energy. In Fig. 1.2 the magnitude of this energy is plotted against the number of molecules. It is clear from this that a wide distribution of energies is present. At 20 °C, the value of RT is 2(293) = 586 cal/mol, but even at this temperature there is a very small fraction of molecules having more than 10,400 cal/mol, which is sufficient to allow them to escape from the surface into the vapor phase. At this temperature the saturated vapor pressure is 1.75 cm Hg. If the temperature is raised to 100 °C, with a value of RT of 746 cal/mol, the distribution curve is moved to the right and a much larger fraction of molecules has the escape energy, which results in a higher saturated

Saturated Vapor Pressure   3

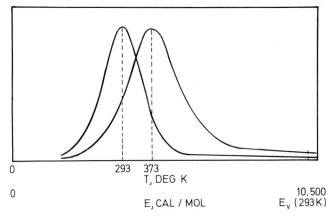

Fig. 1.2. Distribution of the kinetic energies of water molecules at 20° and 100 °C showing the relative number with sufficient energy (10,400 cal/mol) to escape from the surface and become vapor molecules

Table 1.1. Saturated vapor pressure and absolute humidities at saturation

Temp.		p_0, cm Hg	AH_0, g/m³
°C	°F		
− 60	− 94.0	0.0008	
− 40	− 40.0	0.0097	
− 20	− 4.0	0.078	0.92
− 10	14.0	0.21	2.15
− 5	23.0	0.32	3.26
0	32.0	0.46	4.85
5	41.0	0.65	6.80
10	50.0	0.92	9.41
12	53.6	1.05	10.7
14	57.2	1.20	12.1
16	60.8	1.36	13.6
18	64.4	1.55	15.4
20	68.0	1.75	17.3
22	71.6	1.98	19.4
24	75.2	2.24	21.8
26	78.8	2.52	24.4
30	86.0	3.18	30.4
35	95.0	4.22	39.6
40	104.0	5.53	51.1
50	122.0	9.25	83.2
60	140.0	14.92	131
70	158.0	23.37	198
80	176.0	35.51	294
90	184.0	52.58	424
100	212.0	76.00	598
110	230.0	107.46	827
120	248.0	148.91	1,122
140	284.0	271.09	1,968
160	320.0	463.60	3,265

Table 1.2. Saturated vapor pressure between 0 °C and 200 °C (Weast 1981–82 Handbook of Chemistry and Physics)

Temp.		p_0 cm Hg	Temp.		p_0 cm Hg	Temp.		p_0 lb/in²	Temp.		p_0 lb/in²
°C	°F		°C	°F		°C	°F		°C	°F	
0	32.0	0.458	52	125.6	10.21	100	212.0	14.70	152	305.6	72.83
2	35.6	0.529	54	129.2	11.25	102	215.6	15.78	154	309.4	76.77
4	39.2	0.610	56	132.8	12.38	104	219.2	16.92	156	312.8	80.90
6	42.8	0.701	58	136.4	13.61	106	222.8	18.14	158	316.4	85.18
8	46.4	0.805	60	140.0	14.94	108	226.4	19.42	160	320.0	89.65
10	50.0	0.921	62	143.6	16.38	110	230.0	20.78	162	323.6	94.30
12	53.6	1.052	64	147.2	17.93	112	233.6	22.21	164	327.2	99.14
14	57.2	1.200	66	150.8	19.61	114	237.2	23.73	166	330.8	104.2
16	60.8	1.363	68	154.4	21.42	116	240.8	25.33	168	334.4	109.4
18	64.4	1.548	70	158.0	23.37	118	244.4	27.02	170	338.0	114.9
20	68.0	1.754	72	161.6	25.46	120	248.0	28.80	172	341.6	120.5
22	71.6	1.983	74	165.2	27.72	122	251.6	30.67	174	345.2	126.4
24	75.2	2.238	76	168.8	30.14	124	255.2	32.64	176	348.8	132.5
26	78.8	2.521	78	172.4	32.73	126	258.8	34.71	178	352.4	138.8
28	82.4	2.835	80	176.0	35.51	128	262.4	36.89	180	356.0	145.4
30	86.0	3.182	82	179.6	38.49	130	266.0	39.18	182	359.6	152.2
32	89.6	3.566	84	183.2	41.68	132	269.6	41.58	184	363.2	159.3
34	93.2	3.990	86	186.8	45.09	134	273.2	44.10	186	366.8	166.6
36	96.8	4.456	88	190.4	48.71	136	276.8	46.72	188	370.4	174.2
38	100.4	4.969	90	194.0	52.58	138	280.4	49.52	190	374.0	182.0
40	104.0	5.532	92	197.6	56.70	140	284.0	52.42	192	377.6	190.1
42	107.6	6.150	94	201.2	61.09	142	287.6	55.45	194	381.2	198.5
44	111.2	6.826	96	204.8	65.76	144	291.2	58.62	196	384.8	207.2
46	114.8	7.565	98	208.4	70.73	146	294.8	61.94	198	388.4	216.2
48	118.4	8.371	100	212.0	76.00	148	298.4	65.41	200	392.0	225.4
50	122.0	9.251	102	215.6	81.59	150	302.0	69.04			

vapor pressure of 76 cm Hg. Although the fraction of activated molecules is still very small, boiling is possible if the atmospheric pressure is below 76 cm Hg. Values of saturated vapor pressure are presented in Tables 1.1 and 1.2. The latter table is more detailed, giving values of p_0 in cm Hg up to atmospheric pressure and in pounds per square inch above this pressure, which is equivalent to the absolute pressure of saturated steam.

The vapor content of the air can be expressed in several ways other than by partial vapor pressure. The *absolute humidity* (AH) is defined as the mass of water vapor per unit volume of moist air. At saturation, this is called the *saturated absolute humidity* (AH_0). Table 1.1 gives values corresponding to various temperatures. The vapor content of the air may also be expressed by the *moisture content* or *humidity ratio* (W) expressed either in grains of water vapor per lb of dry air, lb water vapor per lb of dry air or g water vapor per kg of dry air. In the psychrometric charts (Figs. 1.3 to 1.5) values of moisture content are plotted along the vertical axes.

Absolute humidity may be calculated from the partial vapor pressure by assuming air and water vapor to be ideal gases obeying Dalton's law, and using the standard molar volume of 0.0224 m³.

$$AH = \frac{p(18 \text{ g/mol})(273 \text{ K})}{76 \text{ cm Hg }(0.0224 \text{ m}^3/\text{mol})T},$$

or, in simplified form,

$$AH = (2{,}887 \text{ g K/cm Hg m}^3)(p/T), \tag{1.3}$$

where p = partial vapor pressure, cm Hg, T = temperature of air, K.

As an *example*, calculate AH_0 at 40 °C from the saturated vapor pressure, p_0.

$$AH_0 = \frac{2887(5.53)}{313} = 51.0 \text{ g/m}^3.$$

This is in close agreement with the value of 51.1 g/m³ given in Table 1.1.

The moisture content (W) is equal to the mass of moisture per unit mass of dry air. It may be calculated from the absolute humidity by dividing it by the density of dry air under standard conditions (1.293 kg/m³). Corrections must be made for temperature and for the partial vapor pressure of dry air.

$$W = \frac{AH(76)T}{(P_a - p)(1.293)(273)},$$

or, in simplified form

$$W = \frac{0.215 \text{ (cm Hg m}^3/\text{g K)}(AH)(T)}{P_a - p}, \tag{1.4}$$

or, by substitution of Eq. (1.3),

$$W = \frac{621 \, p}{P_a - p}, \tag{1.5}$$

where W is expressed in g/kg; divide by 1,000 for lb/lb; multiply by 7 for grain/lb.

Boiling is a form of evaporation which occurs when the saturated vapor pressure becomes equal to the total atmospheric pressure, or the ambient pressure. For example, if the atmospheric pressure is 76 cm Hg, water will boil at a temperature slightly above 100 °C, for at this temperature vapor bubbles have sufficient pressure to remain stable. The slightly higher temperature is required to overcome the hydrostatic pressure due to the depth below the surface and to overcome the surface tension force which tends to collapse the bubble.

1.3 Relative Humidity

The *relative humidity* (H) is defined as the ratio of the partial vapor pressure in the air to the saturated vapor pressure, expressed as a percent.

$$H = p/p_0 \times 100\%. \tag{1.6}$$

Alternatively, the relative humidity is equal to the ratio of the absolute humidity of the air to the absolute humidity at saturation at the same temperature, also expressed as a percent.

$$H = AH/AH_0 \times 100\%. \tag{1.7}$$

The relative humidity may also be calculated from the moisture content as

$$H = \frac{W(P_a - p)}{W_0(P_a - p_0)}, \tag{1.8}$$

where W_0 = moisture content at saturation.

Since the term $(P_a-p)/(P_a-p_0)$ in Eq. (1.8) is slightly greater than unity near room temperature, this equation may be simplified to the approximation

$$H \approx W/W_0 \times 100\%. \tag{1.9}$$

It can be shown that Eqs. (1.6), (1.7), and (1.8) will give identical values of H when calculated from the partial vapor pressure or absolute humidity at the ambient temperature. However, it should be pointed out that different results will be obtained when H is calculated from partial vapor pressures and absolute humidities at the dry-bulb and dew-point temperatures using Eqs. (1.6) and (1.7). The result obtained from Eqs. (1.6) and (1.8) will be correct because a decrease in the temperature will not change the partial vapor pressure, provided the air is still under atmospheric pressure, nor will it change the moisture content of a mass of air. In the use of Eq. (1.7), the AH determined at the lower dew-point temperature from Table 1.1 will be less than that actually existing at the dry-bulb temperature because of the inverse temperature dependency indicated by Eq. (1.3). Therefore, the calculated relative humidity will be high. It is evident that the use of the approximate Eq. (1.9) will result in low values of relative humidity, particularly at higher temperatures where p_0 approaches the magnitude of P_a.

Equation (1.7) may be modified to correct the saturated absolute humidity at the dew point to the actual value at the dry-bulb temperature.

$$H = \frac{AH_{dp} T_{dp}}{AH_0 T} \times 100\% \tag{1.10}$$

corrected for calculation from dew point,

where T_{dp} = dew-point temperature, T = dry-bulb temperature corresponding with AH_0, AH_{dp} = absolute humidity at saturation at T_{dp}.

There is a limitation to the maximum possible relative humidity which can be maintained in a dry kiln operating above the boiling point. Under these circumstances the value of p cannot exceed one atmosphere. Thus:

$$H_{max} = \frac{P_a}{p_0} \times 100\%. \tag{1.11}$$

For *example*, if a dry kiln is operated at 120 °C or 248 °F, $p_0 = 28.80$ lb/in². Assuming standard atmospheric pressure,

$$H_{max} = \frac{14.7}{28.80} \times 100\% = 51.0\%.$$

1.3.1 Use of the Psychrometric Chart

The relationships among dry-bulb, wet-bulb, and dew-point temperatures, relative humidity, specific volume and moisture content of the air are all brought together on the psychrometric charts illustrated in Figs. 1.3, 1.4, and 1.5. These are based on standard atmospheric pressure of 76 cm Hg. Essentially the charts are plots of air moisture content on the vertical axis vs dry-bulb temperature on the horizontal axis. Curved lines of constant relative humidity are given from 10% to 100% of saturation. Lines representing wet-bulb temperature are approximately 30° to the horizontal. The lines of constant moisture content and dew point are horizontal. The state point of the moist air may be located at the intersection of the dry-bulb and wet-bulb or dew-point lines. The relative humidity may then be read from the curved lines and the moisture content (W) from the vertical axis. The dew point is read by following the line of constant moisture content from the state point to the saturation line and reading the dew point from the wet-bulb scale. For *example*, if the dry bulb is 20 °C and the wet bulb is 10 °C, the relative humidity is 25%, the moisture content is 3.5 g/kg, the specific volume is 0.835 m^3/kg dry air, and the dew point is -0.5 °C. The moisture content at saturation at 20 °C is 14.7 g/kg. It is clear from this that the psychrometric chart contains a wealth of related information regarding the state of humid air and may be used as a means of calculating relative humidity from dry-bulb and wet-bulb or dew-point data. Details regarding the derivation of the psychrometric chart and described by Carpenter (1982).

1.3.2 Measurement of Relative Humidity

There are many ways of measuring the relative humidity of the air (de Yong 1982). Two simple and practical methods will be discussed here. These are the sling psychrometer or wet- and dry-bulb thermometer and the dew-point methods.

The *sling psychrometer* consists of two thermometers: one with a dry bulb for measuring the ambient room temperature and the other with a wet bulb covered by a wick dipped in distilled water. Air circulation of at least 3 m/s must be achieved by manual rotation of the instrument or by a fan blowing on a stationary wet bulb, resulting in evaporative cooling. When the wet bulb reaches equilibrium, consisting of saturation of the air surrounding the bulb, with no gain or loss of heat from any other source, the temperature is called the thermodynamic wet-bulb temperature. In an actual situation, there is usually a small amount of heat gained by conduction and radiation from the warmer surroundings which are at the dry-bulb temperature, resulting in a reading slightly higher than the thermodynamic temperature. This error is usually negligible, however, with adequate air circulation. A very large positive error can result if the wick becomes contaminated with deposits from tap water or dirt. This prevents adequate air circulation and the wet-bulb reading will approach the dry-bulb temperature. A high negative error can result in a wet-bulb sensor used in a dry kiln if the water source for the wet bulb is flowing too rapidly, causing cooling of the wet bulb to a temperature approaching that of the water source.

Fig. 1.3. Psychrometric chart, normal temperatures, British units, applicable at a pressure of 76 cm Hg

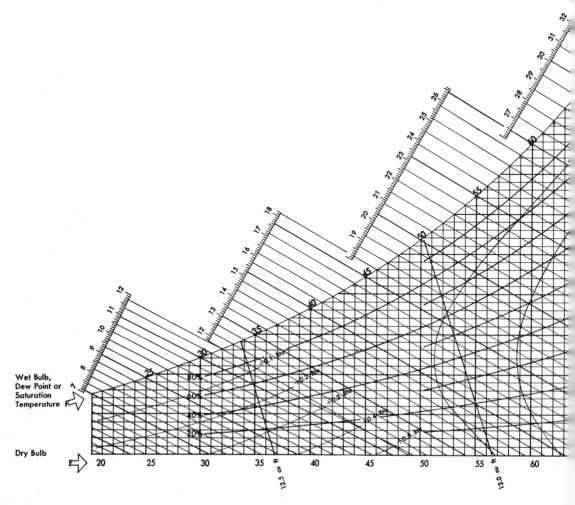

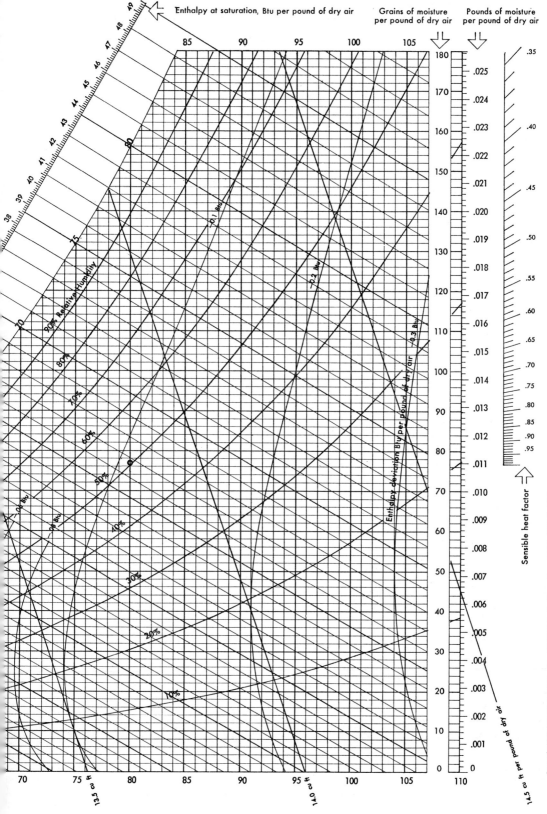

The relative humidity may be read from the psychrometric charts at the intersection of the dry-bulb and wet-bulb lines, provided the observer is near sea level. Other charts are available from Carrier Corporation for higher elevations, and these reveal that the relative humidity corresponding with a given dry-bulb temperature and wet-bulb depression increases significantly with increased altitude and the resulting lower barometric pressure. The effect of reduced barometric pressure can be taken into account by calculation of the relative humidity from Carrier's equation. This is an empirical relationship between the partial vapor pressure, the wet-bulb depression, and the barometric pressure, presented by Carrier (1911) and which has been slightly revised by Carpenter (1982). This equation is very useful for accurate calculation of relative humidity from wet and dry-bulb temperatures and is recommended for barometric pressures which deviate significantly from 76 cm Hg.

$$p = p_{ow} - \frac{(P_a - p_{ow})(F - F_w)}{2{,}830 - 1.44\,F_w} \tag{1.12a}$$

(Carrier's eq., Fahrenheit temperatures)

or,

$$p = p_{ow} - \frac{(P_a - p_{ow})(C - C_w)}{1{,}546 - 1.44\,C_w}, \tag{1.12b}$$

(Carrier's eq., Celsius temperatures),

where p_{ow} = saturated vapor pressure at the wet-bulb temperature, F = dry-bulb temperature, °F, F_w = wet-bulb temperature, °F, C = dry-bulb temperature, °C, C_w = wet-bulb temperature, °C.

Following the calculation of the partial vapor pressure from Eq. (1.12), the relative humidity may be determined from Eq. (1.6).

An approximate relationship between the average barometric pressure and altitude is revealed in Table 1.3 in which a value of 76 cm Hg is assumed at sea level.

Table 1.3. Variation in the average barometric pressure with altitude

Altitude m	Altitude ft	P_a, cm Hg
Sea level	Sea level	76.0
300	1,000	73.3
600	2,000	70.7
900	3,000	68.1
1,200	4,000	65.6
1,500	5,000	63.2
1,800	6,000	60.9
2,100	7,000	58.6
2,400	8,000	56.4
2,700	9,000	54.3
3,000	10,000	52.1
4,600	15,000	42.9
6,100	20,000	34.9
9,100	30,000	22.6

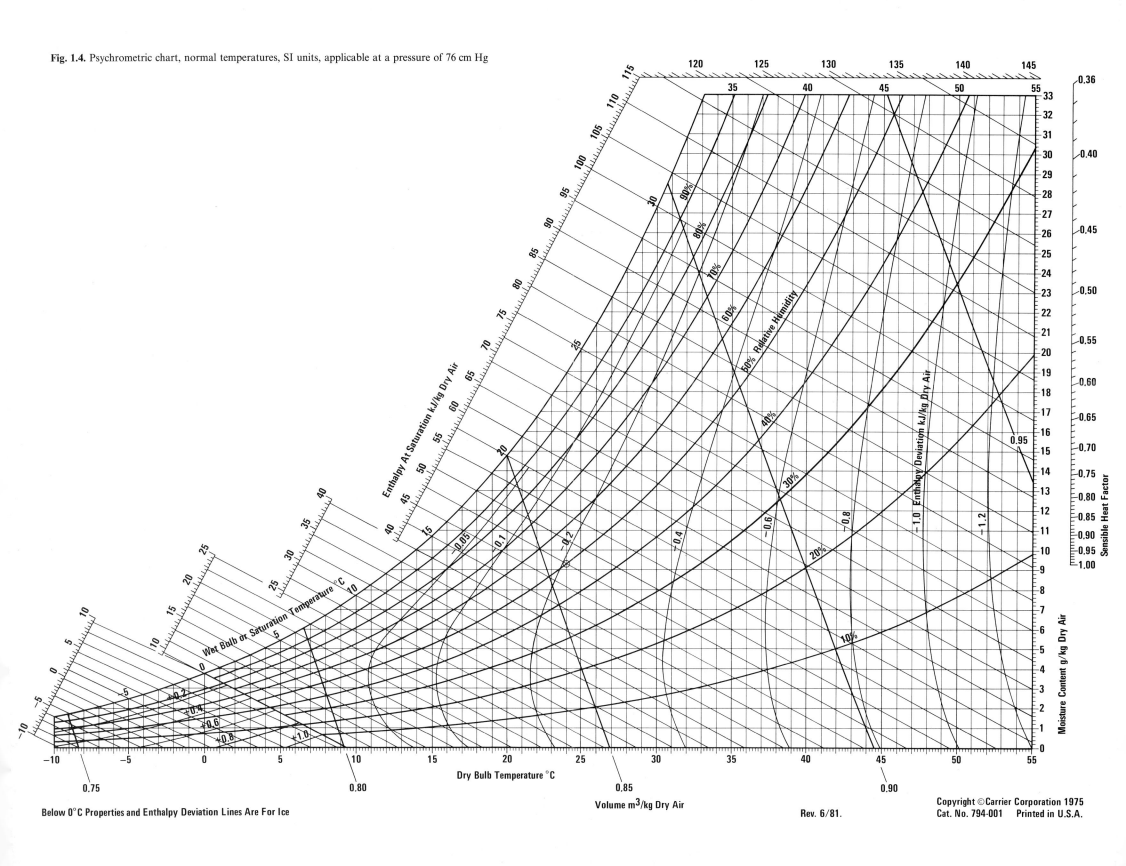

Fig. 1.4. Psychrometric chart, normal temperatures, SI units, applicable at a pressure of 76 cm Hg

Below 0°C Properties and Enthalpy Deviation Lines Are For Ice

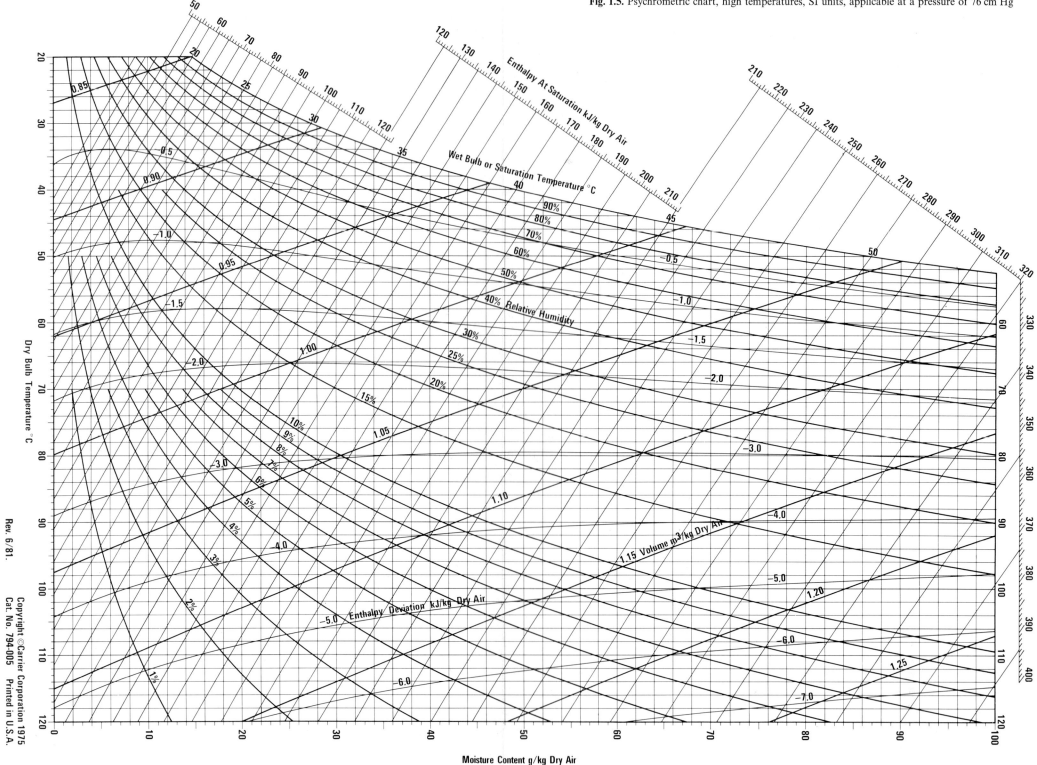

Fig. 1.5. Psychrometric chart, high temperatures, SI units, applicable at a pressure of 76 cm Hg

It is evident from this that the barometric pressure decreases approximately 8 cm for each 1,000 m rise up to 3,000 m. It is best to use the uncorrected barometric pressure (not reduced to sea level) in Eq. (1.12), but a good approximation may be obtained by using the barometric pressure interpolated from Table 1.3 if the former is not available.

The relative humidity may also be determined from the dry-bulb temperature and wet-bulb depression from Tables 1.4 (°F) and 1.5 (°C). These tables are only usable near sea level and are less accurate than the psychrometric charts or Carrier's equation. Tables 1.4 and 1.5 include the range between the freezing and boiling points. The sling psychrometer may also be used below the freezing point; a suitable table is available in the *Handbook of Chemistry and Physics* (Weast 1981–1982). These three methods of determining relative humidity from wet- and dry-bulb data are illustrated by the following.

Example Determination of H from Wet- and Dry-Bulb Temperatures

Determine H at sea level and at an altitude of 750 m when the dry-bulb temperature is 20 °C (68 °F) and the wet-bulb, 10 °C (50 °F).

a) From psychrometric charts, sea level only,
   read H = **25%** from Fig. 1.3
   read H = **25%** from Fig. 1.4.
b) From Carrier's equation,
   read $p_{ow} = 0.92$ cm Hg, $p_0 = 1.75$ cm Hg from Table 1.1.
Substituting into Eq. (1.12a),

$$p = 0.92 - \frac{(76.0 - 0.92)(18)}{2{,}830 - 1.44(50)} = 0.43 \text{ cm Hg}$$

$H = 0.43/1.75 \times 100\% = \mathbf{24.5\%}$, sea level.

An identical value is obtained from Eq. (1.12b) using Celsius temperatures. At 750 m, interpolate $P_a = 69.4$ cm Hg from Table 1.1. Then,

$$p = 0.92 - \frac{(69.4 - 0.92)(18)}{2{,}830 - 1.44(50)} = 0.47 \text{ cm Hg}$$

$H = 0.47/1.75 \times 100\% = \mathbf{27.0\%}$ at 750 m.

c) From Tables 1.6 and 1.7, sea level only
   interpolate H = **23%** from Table 1.4,
   read H = **23%** from Table 1.5.

The *dew point* is the temperature at which the saturated vapor pressure is equal to the existing vapor pressure. The dew point may be measured by gradually cooling the contents of a beaker with a brightly polished surface and carefully observing the temperature at which condensation begins to form on the surface. This is the temperature of appearance of condensation. The contents are then allowed to warm slowly and the temperature of disappearance of the condensation is observed, which is a slightly higher temperature. The dew point is then taken as the average of the two temperatures. Cooling of the beaker may be done with ice added to water if the dew point is above the freezing point, and with dry ice added to isopropanol for temperatures below the freezing point. When the dew point is so determined, the saturated vapor pressure or absolute humidity at the dew point from Table 1.1 or 1.2 is taken as the vapor pressure or absolute humidity of the air and

Table 1.4. Relative humidity[1] and equilibrium moisture content[2] corresponding to various wet-bulb depressions in degrees fahrenheit at sea level (Courtesy of U.S.D.A. Forest Service)

Temperature dry-bulb (°F)		Wet-bulb depression (°F)																																	
		1	2	3	4	5	6	7	8	9	10	11	12	13	14	15	16	17	18	19	20	21	22	23	24	25	26	27	28	29	30	32	34	36	38	40	45	50
30	RH	89	78	67	57	46	36	27	17	6	—	—	—	—	—	—	—	—	—	—	—	—	—	—	—	—	—	—	—	—	—	—	—	—	—	—	—	—
	EMC	—	15.9	12.9	10.8	9.0	7.4	5.7	3.9	1.6																												
35	RH	90	81	72	63	54	45	37	28	19	11	3	—	—	—	—	—	—	—	—	—	—	—	—	—	—	—	—	—	—	—	—	—	—	—	—	—	—
	EMC	—	16.8	13.9	11.9	10.3	8.8	7.4	6.0	4.5	2.9	0.8																										
40	RH	92	83	75	68	60	52	45	37	29	22	15	8	—	—	—	—	—	—	—	—	—	—	—	—	—	—	—	—	—	—	—	—	—	—	—	—	—
	EMC	—	17.6	14.8	12.9	11.2	9.9	8.6	7.4	6.2	5.0	3.5	1.9																									
45	RH	93	85	78	72	64	58	51	44	37	31	25	19	12	6	—	—	—	—	—	—	—	—	—	—	—	—	—	—	—	—	—	—	—	—	—	—	—
	EMC	—	18.3	15.6	13.7	12.0	10.7	9.5	8.5	7.5	6.5	5.3	4.2	2.9	1.5																							
50	RH	93	86	80	74	68	62	56	50	44	38	32	27	21	16	10	5	—	—	—	—	—	—	—	—	—	—	—	—	—	—	—	—	—	—	—	—	—
	EMC	—	19.0	16.3	14.4	12.7	11.5	10.3	9.4	8.5	7.6	6.7	5.7	4.8	3.9	2.8	1.5																					
55	RH	94	88	82	76	70	65	60	54	49	44	39	34	28	24	19	14	9	5	—	—	—	—	—	—	—	—	—	—	—	—	—	—	—	—	—	—	—
	EMC	—	19.5	16.9	15.1	13.4	12.2	11.0	10.1	9.3	8.4	7.6	6.8	6.0	5.3	4.5	3.6	2.5	1.3																			
60	RH	94	89	83	78	73	68	63	58	53	48	43	39	34	30	26	21	17	13	9	5	1	—	—	—	—	—	—	—	—	—	—	—	—	—	—	—	—
	EMC	—	19.9	17.4	15.6	13.9	12.7	11.6	10.7	9.9	9.1	8.3	7.6	6.9	6.3	5.6	4.9	4.1	3.2	2.3	1.3	0.2																
65	RH	95	90	84	80	75	70	66	61	56	52	48	44	39	36	32	27	24	20	16	13	8	6	2	—	—	—	—	—	—	—	—	—	—	—	—	—	—
	EMC	—	20.3	17.8	16.1	14.4	13.3	12.1	11.2	10.4	9.7	9.1	8.3	7.7	7.1	6.5	5.8	5.2	4.5	3.8	3.0	2.3	1.4	0.4														
70	RH	95	90	86	81	77	72	68	64	59	55	51	48	44	40	36	33	29	25	22	19	15	12	9	6	3	—	—	—	—	—	—	—	—	—	—	—	—
	EMC	—	20.6	18.2	16.5	14.9	13.7	12.5	11.6	10.9	10.1	9.4	8.8	8.3	7.7	7.2	6.6	6.0	5.5	4.9	4.3	3.7	2.9	2.3	1.5	0.7												
75	RH	95	91	86	82	78	74	70	66	62	58	54	51	47	44	41	37	34	31	28	24	21	18	15	12	10	7	4	1	—	—	—	—	—	—	—	—	—
	EMC	—	20.9	18.5	16.8	15.2	14.0	12.9	12.0	11.2	10.5	9.8	9.3	8.7	8.2	7.7	7.2	6.7	6.2	5.6	5.1	4.7	4.1	3.5	2.9	2.3	1.7	0.9	0.2									
80	RH	96	91	87	83	79	75	72	68	64	61	57	54	50	47	44	41	38	35	32	29	26	23	20	18	15	12	10	7	5	3	—	—	—	—	—	—	—
	EMC	—	21.0	18.7	17.0	15.5	14.3	13.2	12.3	11.5	10.9	10.1	9.7	9.1	8.6	8.1	7.7	7.2	6.8	6.3	5.8	5.4	5.0	4.5	4.0	3.5	3.0	2.4	1.8	1.1	0.3							
85	RH	96	92	88	84	80	76	73	70	66	63	59	56	53	50	47	44	41	38	36	33	30	28	25	23	20	18	15	13	11	9	4	—	—	—	—	—	—
	EMC	—	21.2	18.8	17.2	15.7	14.5	13.5	12.5	11.8	11.2	10.5	10.0	9.5	9.0	8.5	8.1	7.6	7.2	6.7	6.3	5.8	5.4	5.2	4.8	4.3	3.4	3.4	3.0	2.4	1.7	0.9						
90	RH	96	92	89	85	81	78	74	71	68	65	61	58	55	52	49	47	44	41	39	36	34	31	29	26	24	22	19	17	15	13	9	5	1	—	—	—	—
	EMC	—	21.3	18.9	17.3	15.9	14.7	13.7	12.8	12.0	11.4	10.7	10.2	9.7	9.3	8.8	8.4	8.0	7.6	7.2	6.8	6.5	6.1	5.7	5.3	4.9	4.6	4.2	3.8	3.3	2.8	2.1	1.3	0.4				
95	RH	96	92	89	85	82	79	75	72	69	66	63	60	57	55	52	49	46	44	42	39	37	34	32	30	28	26	23	22	20	17	14	10	6	2	—	—	—
	EMC	—	21.3	19.0	17.4	16.1	14.9	13.9	12.9	12.0	11.6	11.0	10.5	10.0	9.5	9.1	8.7	8.2	7.9	7.5	7.1	6.8	6.4	6.1	5.7	5.3	5.1	4.8	4.4	4.0	3.6	3.0	2.3	1.5	0.6			
100	RH	96	93	89	86	83	80	77	73	70	68	65	62	59	56	54	51	49	46	44	41	39	37	35	33	30	28	26	24	22	21	17	13	10	7	4	—	—
	EMC	—	21.3	19.0	17.5	16.1	15.0	13.9	13.1	12.4	11.8	11.2	10.6	10.1	9.6	9.2	8.9	8.5	8.1	7.8	7.4	7.1	6.8	6.4	6.1	5.7	5.4	5.2	4.9	4.6	4.2	3.6	3.1	2.4	1.6	0.7		
105	RH	96	93	90	87	83	80	77	74	71	69	66	63	60	58	55	53	50	48	46	44	42	40	37	35	34	31	29	28	26	24	20	17	14	11	8	—	—
	EMC	—	21.4	19.0	17.5	16.2	15.1	14.0	13.2	12.5	11.9	11.3	10.8	10.3	9.8	9.4	9.0	8.7	8.3	7.9	7.6	7.3	6.9	6.7	6.4	6.1	5.7	5.4	5.2	4.8	4.6	4.2	3.6	3.1	2.4	1.8		
110	RH	97	93	90	87	84	81	78	75	73	70	67	65	62	60	57	55	52	50	48	46	44	42	40	38	36	34	32	30	28	26	23	20	17	14	11	4	—
	EMC	—	21.4	19.0	17.5	16.2	15.1	14.1	13.3	12.6	12.0	11.4	10.8	10.4	9.9	9.5	9.2	8.8	8.4	8.1	7.7	7.5	7.2	6.8	6.6	6.3	6.0	5.7	5.4	5.2	4.8	4.5	4.0	3.5	3.0	2.5	1.1	

14 Basic Wood–Moisture Relationships

Measurement of Relative Humidity

	97	93	90	88	85	82	79	76	74	71	68	66	63	61	58	56	54	52	50	48	45	43	41	40	38	36	34	32	31	29	26	23	20	17	14	8	2
115	–	21.4	19.0	17.5	16.2	15.1	14.1	13.4	12.7	12.1	11.5	10.9	10.4	10.0	9.6	9.3	8.9	8.6	8.2	7.8	7.6	7.3	7.0	6.7	6.5	6.2	5.9	5.6	5.4	5.2	4.7	4.3	3.9	3.4	2.9	1.7	0.4
120	97	94	91	88	85	82	80	77	74	72	69	67	65	62	60	58	55	53	51	49	47	45	43	41	40	38	36	34	33	31	28	25	22	19	17	10	5
	–	21.3	19.0	17.4	16.2	15.1	14.1	13.4	12.7	12.1	11.5	11.0	10.5	10.0	9.7	9.4	9.0	8.7	8.3	7.9	7.7	7.4	7.2	6.8	6.6	6.3	6.1	5.8	5.6	5.4	5.0	4.6	4.2	3.7	3.3	2.3	1.1
125	97	94	91	88	86	83	80	77	75	73	70	68	65	63	61	59	57	55	53	51	48	47	45	43	41	39	38	36	35	33	30	27	24	22	19	13	8
	–	21.2	18.9	17.3	16.1	15.0	14.0	13.4	12.7	12.1	11.5	11.0	10.5	10.0	9.7	9.4	9.0	8.7	8.3	8.0	7.7	7.5	7.2	7.0	6.7	6.5	6.2	6.0	5.8	5.5	5.2	4.8	4.4	4.0	3.6	2.7	1.6
130	97	94	91	89	86	83	81	78	76	73	71	69	67	64	62	60	58	56	54	52	50	48	47	45	43	41	40	38	37	35	32	29	26	24	21	15	10
	–	21.0	18.8	17.2	16.0	14.9	14.0	13.4	12.7	12.1	11.5	11.0	10.5	10.0	9.7	9.4	9.0	8.7	8.3	8.0	7.8	7.6	7.3	7.0	6.8	6.6	6.4	6.1	5.9	5.6	5.3	4.9	4.6	4.2	3.8	3.0	2.0
140	97	95	92	89	87	84	82	79	77	75	73	70	68	66	64	62	60	58	56	54	53	51	49	47	46	44	43	41	40	38	35	32	30	27	25	19	14
	–	20.7	18.6	16.9	15.8	14.8	13.8	13.2	12.5	11.9	11.4	10.9	10.4	10.0	9.6	9.4	9.0	8.7	8.4	8.0	7.8	7.6	7.3	7.1	6.9	6.6	6.4	6.2	6.0	5.8	5.4	5.1	4.8	4.4	4.1	3.4	2.6
150	98	95	92	90	87	85	82	80	78	76	74	72	70	68	66	64	62	60	58	57	55	53	51	49	48	46	45	43	42	41	38	36	33	30	28	23	18
	–	20.2	18.4	16.6	15.4	14.5	13.7	13.0	12.4	11.8	11.2	10.8	10.3	9.9	9.5	9.2	8.9	8.6	8.3	8.0	7.8	7.5	7.3	7.1	6.9	6.7	6.4	6.2	6.0	5.8	5.4	5.2	4.9	4.5	4.2	3.6	2.9
160	98	95	93	90	88	86	83	81	79	77	75	73	71	69	67	65	64	62	60	58	57	55	53	52	50	49	47	46	44	43	41	38	35	33	31	25	21
	–	19.8	18.1	16.2	15.2	14.2	13.4	12.7	12.1	11.5	11.0	10.6	10.1	9.7	9.4	9.1	8.8	8.5	8.2	7.9	7.7	7.4	7.2	7.0	6.8	6.7	6.4	6.2	6.0	5.8	5.5	5.2	4.9	4.6	4.3	3.7	3.2
170	98	95	93	91	89	86	84	82	80	78	76	74	72	70	69	67	65	63	62	60	59	57	55	53	52	51	49	48	47	45	43	40	38	35	33	28	24
	–	19.4	17.7	15.8	14.8	13.9	13.2	12.4	11.8	11.3	10.8	10.4	9.9	9.6	9.2	9.0	8.6	8.4	8.0	7.9	7.6	7.4	7.2	6.9	6.7	6.6	6.4	6.2	6.0	5.7	5.5	5.2	4.9	4.6	4.4	3.7	3.2
180	98	96	94	91	89	87	85	83	81	79	77	75	73	72	70	68	67	65	63	62	60	58	57	55	54	52	51	50	48	45	42	40	38	35	30	26
	–	18.9	17.3	15.5	14.5	13.7	12.9	12.2	11.6	11.1	10.6	10.1	9.7	9.4	9.0	8.8	8.4	8.1	7.8	7.6	7.4	7.2	7.0	6.8	6.6	6.5	6.4	6.2	6.0	5.8	5.4	5.2	4.8	4.6	4.4	3.8	3.3
190	98	96	94	92	90	88	85	84	82	80	78	76	75	73	71	69	68	66	65	63	62	60	58	57	56	54	53	51	50	49	46	44	42	39	37	32	28
	–	18.5	16.9	15.2	14.2	13.4	12.7	12.0	11.4	10.9	10.5	10.1	9.6	9.4	9.0	8.8	8.4	8.1	7.7	7.6	7.2	7.0	6.8	6.6	6.4	6.2	6.0	5.9	5.7	5.5	5.3	5.0	4.8	4.5	4.4	3.8	3.3
200	98	96	94	92	90	88	86	84	82	80	79	77	75	74	72	70	69	67	66	64	63	61	60	58	57	55	54	52	51	48	46	43	41	39	34	30
	–	18.1	16.4	14.9	14.0	13.2	12.4	11.8	11.2	10.8	10.3	9.8	9.4	9.1	8.8	8.4	8.1	7.7	7.5	7.2	7.0	6.9	6.6	6.4	6.2	6.0	5.9	5.7	5.6	5.4	5.2	4.9	4.7	4.5	4.3	3.8	3.3
210	98	96	94	92	90	88	86	85	83	81	79	78	76	75	73	71	70	68	67	65	64	63	61	60	59	57	56	53	52	50	47	45	43	41	36	32
	–	17.7	16.0	14.6	13.8	13.0	12.2	11.7	11.1	10.6	10.0	9.7	9.2	9.0	8.7	8.3	8.0	7.6	7.4	7.1	6.9	6.8	6.5	6.3	6.1	5.9	5.8	5.5	5.4	5.3	5.1	4.8	4.6	4.4	4.2	3.7	3.2

[1] Relative-humidity values in roman type.
[2] Equilibrium-moisture-content values in italic type.
Z M 87885 F

Table 1.5. Relative humidity corresponding to various wet-bulb depressions in degrees Celsius at sea level (Courtesy of Blue M Electric Co.)

Dry bulb °C	Wet-bulb depression (°C)																											
	0.5	1.0	1.5	2.0	2.5	3.0	3.5	4.0	4.5	5	6	7	8	9	10	11	12	13	14	15	16	18	20	22	24	26	28	30
2	92	83	75	67	59	52	43	36	27	20																		
4	93	85	77	70	63	56	48	41	34	28	15																	
6	94	87	80	73	66	60	54	47	41	35	23	11																
8	94	87	81	74	68	62	56	50	45	39	28	17																
10	94	88	82	76	71	65	60	54	49	44	34	23	14															
12	94	89	84	78	73	68	63	58	53	48	38	30	21	12	4													
14	95	90	84	79	74	69	65	60	55	51	41	33	24	16	10	7												
16	95	90	85	80	76	71	67	62	58	54	45	37	29	21	14	13	6											
18	95	90	86	81	76	73	69	65	61	57	49	42	35	27	20	17	11											
20	96	91	87	82	78	74	70	66	62	58	51	44	36	30	23													
22	96	92	87	83	79	75	72	68	64	60	53	46	40	34	27	21	16	11										
24	96	92	88	85	81	77	74	70	66	63	56	49	43	37	31	26	21	14	10									
26	96	92	89	85	81	77	74	71	67	64	57	51	45	39	34	28	23	18	13			12						
28	96	92	89	85	81	78	75	72	68	65	59	53	47	42	37	31	27	21	17	13	12	15	10					
30	96	93	89	86	82	79	76	73	70	67	61	55	50	44	39	35	30	24	20	16	15	18	13					
32	96	93	90	86	83	80	77	74	71	68	62	56	51	46	41	36	32	27	23	19	18	21	15	12				
34	97	93	90	87	84	81	77	74	71	69	63	58	53	48	43	38	34	30	26	22	21	23	17	14				
36	97	93	90	87	84	81	78	75	72	70	64	59	54	50	45	41	36	32	28	24	23	25	19	16	12			
38	97	94	90	87	84	81	79	76	73	70	65	60	56	51	46	42	38	34	30	26	25	27	21	18	14			
40	97	94	91	88	85	82	79	76	74	71	66	61	57	52	48	44	40	36	32	29	27	28	23	20	16			
42	97	94	91	88	85	82	80	77	74	72	67	62	58	53	49	45	41	38	34	31	30	30	25	20	16			
44	97	94	91	88	86	83	80	77	75	73	68	63	59	54	50	47	43	39	36	32	31	31	26	21	17	11		
46	97	94	91	89	86	83	80	78	76	73	68	64	60	55	52	48	44	41	37	34	33	32	27	23	19	13		
48	97	94	92	89	86	84	81	78	76	74	69	65	61	56	53	49	45	42	39	35	34	33	29	24	20	15	11	
50	97	94	92	89	87	84	82	79	77	75	70	65	62	57	54	50	47	43	40	37	35	35	30	26	21	16	12	
52	97	94	92	89	87	85	82	79	77	75	70	66	62	58	55	51	48	44	41	38	35	30	25	20	23	18	11	11
54	97	95	92	90	87	85	82	80	78	76	71	67	63	59	56	52	49	45	42	39	36	31	26	21	24	19	12	12
56	97	95	92	90	87	85	83	80	78	76	72	68	64	60	57	53	50	46	43	40	38	32	27	23	25	20	14	13
58	97	95	93	90	88	86	83	80	79	77	72	68	64	61	57	54	51	47	44	42	39	33	29	24	26	22	15	
60	98	95	93	90	88	86	83	81	79	77	73	69	65	62	58	55	52	48	45	43	40	35	30	26	28	23	17	
62	98	95	93	91	88	86	84	81	79	78	73	69	66	62	59	56	53	49	46	43	41	36	31	27	23	19		
64	98	95	93	91	88	86	84	82	80	78	74	70	66	63	59	56	53	50	47	44	42	37	32	28	24	20		

16   Basic Wood–Moisture Relationships

Measurement of Relative Humidity

the relative humidity is calculated from Eqs. (1.6) or (1.10). Alternatively, the relative humidity may be read from the psychrometric chart at the state point of the air which is found at the intersection of the horizontal dew-point line and the vertical dry-bulb line. When this is done it will be noted that the dew point is always below the wet-bulb temperature except when the air is saturated. An advantage of this method is that the result is not affected by the altitude of the observer. The following example will illustrate these methods.

Example Determination of H from Dew Point
  Dry-bulb temperature = 20 °C
  Dew-point temperature = 5 °C.

Determine H from: (a) partial vapor pressure, (b) absolute humidity, and (c) the psychrometric chart.

a) $p = 0.65$ cm Hg, $p_o = 1.75$ cm Hg, Table 1.1
  $H = 0.65/1.75 \times 100\% = \mathbf{37.1\%}$.
b) $AH_{dp} = 6.80$ g/m³, $AH_o = 17.3$ g/m³, Table 1.1
  $T = 293$ K, $T_{dp} = 278$ K
  $H = (6.80/17.3) \cdot (278/293) \times 100\% = \mathbf{37.3\%}$.
c) Read **37%** from Fig. 1.4.

1.3.3 Control of Relative Humidity

The relative humidity within a room or chamber may be controlled by: (a) regulating the dry-bulb temperature by thermostatically operated heating or cooling devices and (b) controlling the dew point or wet-bulb temperature thermostatically to maintain the desired relative humidity. The latter may be done at temperatures above the freezing point by controlling the temperature of a water bath from which water is sprayed into a moving current of dry air. The spray will cause rapid evaporation of water as the dry air moves through it and also evaporative cooling. The air must be reheated to obtain the desired dry-bulb temperature. The amount of water vapor taken up by the air may be increased or decreased by increasing or decreasing the temperature of the water bath until the desired relative humidity is maintained. Humidity may also be regulated by adding live steam to the moving air to increase humidity, or by passing the air stream through refrigerator coils at a temperature below the dew point to reduce the relative humidity. In the case of a dry kiln, the relative humidity is usually controlled by a wet-bulb sensor and controller. Live steam is added to the kiln to increase the wet-bulb temperature, and vents at the top are opened to allow the escape of moist air and thus to decrease the wet-bulb temperature.

Within a confined space, aqueous solutions of salts or sulfuric acid may be used to control relative humidity. Pure water will give saturated conditions or 100% relative humidity. When a solute is added to water, the vapor pressure is reduced in proportion to the mole fraction in the case of dilute solutions where Raoult's law is obeyed. When saturated solutions are used at a fixed temperature, a constant relative humidity is maintained. A list of inorganic salts and their corresponding relative humidities are provided in Table 1.6. The relative humidities which may be maintained over sulfuric acid solutions are presented in Table 1.7.

Table 1.6. Relative humidity and partial vapor pressure over saturated salt solutions at 20 °C (Weast 1981–1982 Handbook of Chemistry and Physics)

Chemical	Formula	H, %	p, cm Hg
Water	H_2O	100	1.75
Copper sulfate	$CuSO_4 \cdot 5H_2O$	98	1.70
Ammonium monophosphate	$NH_4H_2PO_4$	93.1	1.62
Sodium sulfate	$Na_2SO_4 \cdot 10H_2O$	93	1.61
Potassium chromate	K_2CrO_4	88	1.53
Ammonium sulfate	$(NH_4)_2SO_4$	81	1.41
Ammonium chloride	NH_4Cl	79.5	1.38
Sodium acetate	$NaC_2H_3O_2 \cdot 3H_2O$	76	1.32
Sodium nitrite	$NaNO_2$	66	1.15
Sodium bromide	$NaBr \cdot 2H_2O$	58	1.01
Sodium dichromate	$Na_2Cr_2O_7 \cdot 2H_2O$	52	0.903
Potassium nitrite	KNO_2	45	0.781
Chromium trioxide	CrO_3	35	0.608
Calcium chloride	$C_aCl_2 \cdot 6H_2O$	32.5	0.561
Potassium acetate	$KC_2H_3O_2$	20	0.347
Lithium chloride	$L_iCl \cdot H_2O$	15	0.260

Table 1.7. Relative humidity and vapor pressure over sulfuric acid solutions at 20 °C (Weast 1981–1982 Handbook of Chemistry and Physics)

Density of acid solution, g/cm³	H, %	p, cm Hg
1.05	97.5	1.70
1.10	93.9	1.63
1.15	88.8	1.54
1.20	80.5	1.40
1.25	70.4	1.22
1.30	58.3	1.01
1.35	47.2	0.83
1.40	37.1	0.65
1.50	18.8	0.33
1.60	8.5	0.15
1.70	3.2	0.06

1.4 Equilibrium Moisture Content and the Sorption Isotherm

Moisture content (M) is the quantity of moisture in wood, expressed as a percentage of oven-dry mass. The latter is defined as the constant mass obtained after wood has been dried in an air oven maintained at 102 ± 3 °C.

$$M = \frac{w_g - w_o}{w_o} \times 100\%,$$

where w_g = green or moist mass, w_o = oven-dry mass.

The moisture in wood has two forms: bound or hygroscopic water, and free or capillary water. Bound water is found in the cell wall and is believed to be hydrogen bonded to the hydroxyl groups of primarily the cellulose and hemicelluloses, and to a lesser extent also to the hydroxyl groups of lignin. Bound-water moisture content is limited by the number of sorption sites available and by the number of molecules of water which can be held on a sorption site. When only bound water is present an equilibrium exists between the moisture content of the wood and the relative humidity of the surrounding air. The moisture content in equilibrium with a given relative humidity is called the *equilibrium moisture content* (EMC). Although relative humidity is the most important variable affecting EMC, other factors are: mechanical stress, the drying history of the wood, the species and specific gravity of the wood, the extractive content, and the temperature. These other factors are discussed in detail by Skaar (1972). In general, an increase in compressive stress decreases EMC. In this connection, the cellular structure of solid wood exerts restraint such that the EMC of solid wood is expected to be lower than that of thin microtome sections. The EMC of never-dried wood is higher than that of wood that has undergone drying. In addition, the EMC is higher during desorption than during adsorption, with an approximate ratio of 0.8 between the adsorption and desorption values at a given relative humidity. These effects can be explained by: (a) incomplete rehydration of sorption sites during a subsequent adsorption cycle and (b) the effect of compressive stresses during swelling. Increased extractives content usually decreases EMC because their presence bulks the cell wall with materials of relatively low hygroscopicity as shown by a study conducted by Wangaard and Granados (1967). A decrease in specific gravity usually results in an increase in EMC (Feist and Tarkow 1967; Skaar 1972).

Fiber saturation point (FSP) has been defined by Tiemann (1906) as the moisture content at which the cell walls are saturated with bound water with no free water in the lumens. Stamm (1964) has interpreted FSP as the moisture content corresponding with abrupt changes in physical properties of wood such as shrinkage, mechanical strength, electrical conductivity, and heat of wetting with changing moisture content. The values obtained by Stamm for *Picea sitchensis* were between 24% and 31%, with most of them between 27% and 31%. There are considerable variations with species, with values extending from 21% for *Thuja plicata* to 24% for *Sequoia sempervirens* and 32% for *Tilia americana*. Feist and Tarkow (1967) report further that FSP tends to increase as wood specific gravity decreases, possibly due to decreased mechanical restraint in the thin cell walls. Other methods of measurement have resulted in significantly higher values of FSP. These and the broader aspects of fiber saturation point will be discussed in detail in Chap. 4.8. Fiber saturation point appears to have the greatest importance in the wood science field, as it relates to changes in properties. Therefore it is proposed that it be redefined as the moisture content which corresponds with abrupt changes in the physical properties of wood. In this context, an average FSP of 30% may be assumed for most woods growing in the temperate zones. The relationship between EMC and relative humidity at a given temperature is called the *sorption isotherm*, depicted in Fig. 1.6, and with numerical data between the freezing and boiling points presented in Table 1.4. Table 1.8 contains sorption isotherm data above the boiling point extrapolated from the low-temperature data by Simpson and Rosen (1981).

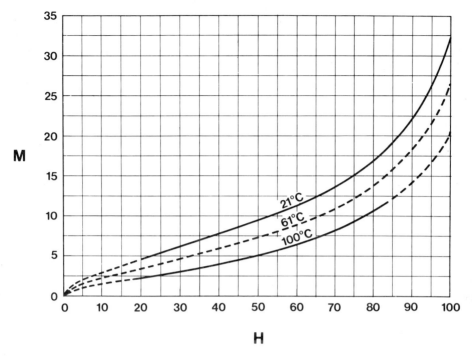

Fig. 1.6. Sorption isotherms of wood at three temperatures from Rasmussen (1961)

Both tables have been compiled by the USDA Forest Service and represent an average for sorption and desorption data representative of several woods. Although significant variations from these values may occur in specific woods as discussed above, these data are extremely useful for many practical applications where the sorption isotherm of a particular wood is not available.

Reference to Fig. 1.6 and Tables 1.4 and 1.8 indicates that the EMC of wood decreases slightly with rising temperature, although the partial water-vapor pressure increases very rapidly, approximately doubling for each 10 °C rise in temperature. Therefore wood moisture content has a much stronger correlation with relative humidity than with absolute humidity. The FSP of wood decreases approximately 0.1% per 1 °C rise in temperature (Fig. 6.3). The decrease in EMC is then approximately proportional to the ratio of the EMC to the FSP. Therfore

$$dM/dT \approx M/M_f \, (0.1\%/°C), \tag{1.14}$$

where M_f = moisture content at FSP, M = EMC.

The free water in wood is in liquid form in the lumens or voids of the wood. The amount of free water which wood may hold is limited by the fractional void space or porosity of the wood. There is no hydrogen bonding and therefore free water is held only by weak capillary forces and cannot cause swelling or shrinking or changes in most other physical properties because the cell wall is already saturated by the much more tightly bound hygroscopic water.

Table 1.8. Relative humidity and equilibrium moisture content corresponding to various wet-bulb depressions at temperatures above the boiling point in degrees Fahrenheit at sea level (Simpson and Rosen 1981)

Wet-bulb depression °F	Dry-bulb temperature																					
	215			220			225			230			235			240			245			
	WB[a] °F	H %	EMC %	WB °F	H %	EMC %	WB °F	H %	EMC %	WB °F	H %	EMC %	WB °F	H %	EMC %	WB °F	H %	EMC %	WB °F	H %	EMC %
4	211	92.1	14.4																		
5	210	90.3	13.5																		
6	209	88.5	12.7																		
7	208	86.7	12.0																		
8	207	85.0	11.4																		
9	206	83.3	10.8	211	84.0	10.8															
10	205	81.6	10.3	210	82.3	10.3															
11	204	79.9	9.8	209	80.6	9.8															
12	203	78.3	9.4	208	79.0	9.4															
13	202	76.7	9.0	207	77.4	9.0															
14	201	75.1	8.7	206	75.9	8.6	211	76.2	8.5												
15	200	73.5	8.3	205	74.3	8.3	210	74.7	8.2												
16	199	72.0	8.0	204	72.8	8.0	209	73.2	7.9												
17	198	70.5	7.8	203	71.3	7.7	208	71.8	7.6												
18	197	69.0	7.5	202	69.9	7.4	207	70.3	7.3												
19	196	67.6	7.3	201	68.4	7.2	206	68.9	7.1	211	69.3	7.0									
20	195	66.1	7.0	200	67.0	7.0	205	67.5	6.9	210	67.9	6.7									
22	193	63.3	6.6	198	64.2	6.6	203	64.7	6.4	208	65.2	6.3									
24	191	60.7	6.2	196	61.5	6.2	201	62.1	6.1	206	62.6	6.0	211	63.1	5.9						
26	189	58.1	5.9	194	59.0	5.8	199	59.5	5.7	204	60.1	5.7	209	60.6	5.6						
28	187	55.5	5.6	192	56.4	5.5	197	57.1	5.4	202	57.6	5.4	207	58.2	5.3						
30	185	53.1	5.3	190	54.0	5.3	195	54.7	5.2	200	55.3	5.1	205	55.9	5.0	210	56.5	4.9			
35	180	47.3	4.7	185	48.3	4.6	190	49.0	4.6	195	49.7	4.5	200	50.3	4.4	205	51.0	4.4	210	51.8	4.3
40	175	42.1	4.2	180	43.1	4.2	185	43.8	4.1	190	44.5	4.0	195	45.2	4.0	200	45.9	3.9	205	46.7	3.8
45	170	37.3	3.8	175	38.3	3.7	180	39.0	3.7	185	39.8	3.6	190	40.5	3.5	195	41.2	3.5	200	42.0	3.4
50	165	32.9	3.4	170	33.9	3.4	175	34.7	3.3	180	35.4	3.3	185	36.2	3.2	190	36.9	3.1	195	37.7	3.1

Equilibrium Moisture Content and the Sorption Isotherm

WB[a]	DB	EMC	ΔT	DB	EMC	ΔT	DB	EMC	ΔT	DB	EMC	ΔT	DB	EMC	ΔT	DB	EMC	ΔT	DB	EMC
55	160	28.9	3.0	165	29.9	3.0	170	30.7	2.9	175	31.4	2.9	180	32.2	2.8	185	33.1	2.8	190	33.8
60	155	25.3	2.7	160	26.2	2.7	165	27.0	2.6	170	27.8	2.6	175	28.6	2.6	180	29.3	2.6	185	30.2
70	145	19.0	2.2	150	19.9	2.2	155	20.7	2.1	160	21.5	2.1	165	22.2	2.1	170	23.0	2.1	175	23.8
80	135	13.8	1.7	140	14.7	1.7	145	15.5	1.7	150	16.2	1.7	155	17.0	1.7	160	17.7	1.7	165	18.4
90	125	9.6	1.3	130	10.4	1.3	135	11.2	1.3	140	11.9	1.3	145	12.6	1.3	150	13.3	1.3	155	14.0
100	115	6.2	0.9	120	7.0	0.9	125	7.7	1.0	130	8.4	1.0	135	9.1	1.0	140	9.7	1.0	145	10.4
110	105	3.5	0.5	110	4.1	0.6	115	4.9	0.6	120	5.5	0.7	125	6.2	0.7	130	6.8	0.7	135	7.4
120	95	1.3	0.2	100	1.9	0.3	105	3.0	0.4	110	3.2	0.4	115	3.8	0.5	120	4.4	0.5	125	5.0
130	85	—	—	90	—	—	95	0.8	0.1	100	1.4	0.2	105	1.9	0.2	110	2.5	0.3	115	2.7
140	75	—	—	80	—	—	85	—	—	90	—	—	95	—	—	100	0.9	0.1	105	1.4

[a] WB = Wet-bulb temperature

1.5 The Effect of Changes in Pressure and Temperature on Relative Humidity

When a mass of air is compressed at constant temperature, the partial vapor pressure is increased in proportion to the increase in the total pressure while the saturated vapor remains unchanged. It is clear from Eq. (1.6) that the relative humidity then increases until p and p_0 become equal when condensation begins. For example, air initially at a relative humidity of 25% will reach saturation when the pressure is increased by a factor of four. Any further increase in pressure is accompanied by condensation. A practical example is the need to periodically remove condensation from a compressed air tank.

When a mass of air is heated at constant pressure, the saturated vapor presssure increases rapidly, approximately doubling with every 10 °C or 18 °F rise in temperature, while the partial vapor pressure remains unchanged. Therefore, the relative humidity will be correspondingly halved with each 10 °C rise in temperature. Such extreme decreases in relative humidity usually do not occur in heated buildings, however, because there are several sources of additional moisture such as vapors given off by cooking, drying processes, or exhaled air of the occupants. These sources raise the absolute humidity, the partial vapor pressure, and the relative humidity, but the increase is usually much less than that in the saturated vapor pressure due to the elevated temperature. Thus the heated interior usually has a higher absolute humidity but a lower relative humidity than the outside. This corresponds to a higher interior dew point. If the interior dew point exceeds the outside temperature, there is a potential problem with condensation on cold exterior surfaces such as windows or uninsulated walls. To prevent such problems, a vapor barrier may be installed on the warm side of a wall to minimize the flow of moist air toward the cold exterior surface. An alternative would be to place a closed-cell insulation on the exterior of a building to keep the walls at a temperature above the interior dew point.

Hoadley (1967) has shown that the outdoor relative humidity in the northeastern United States remains approximately constant at 80%, while the interior value can extend from 10% up to 80% with the maximum during the summer. This can cause changes in the EMC of wood from approximately 2% in winter to 17% in summer. Such extreme changes rarely occur, however, but one may expect that the average moisture content of interior woodwork or furniture could vary from 6% to 12% with the seasons in the northeastern United States. This could cause a volumetric swelling of the wood in the range of 2.5% to 4%, or an average linear swelling of 1.25% to 2% in the transverse directions (see Sect. 1.9).

Annual fluctuations in humidity are very low in moist, warm areas such as the coastal regions of the southeastern United States. In such locations, the relative humidity could be over 80% much of the year corresponding to a relatively high EMC. In desert-like climates of the western United States with their very low humidities, the EMC of interior woodwork should be much lower. It is therefore important that lumber be dried to a moisture content approximately equal to the EMC of its intended use. This topic is discussed in detail in Chap. 14 of the *Wood Handbook* (USDA 1974).

The operation of the drying oven may be explained on a basis of the influence of temperature on relative humidity which, in turn, determines the EMC of wood.

Assume room temperature is 20 °C with a relative humidity of 50%. The value of p_0 is 1.75 cm Hg and the partial vapor pressure, p, is 0.88 cm Hg. At 102 °C, $p_0 = 83$ cm Hg from Eq. (1.2). If the expansion of the air in the oven is neglected, the relative humidity will be $(0.88/83) \times 100\% = 1.1\%$. Reference to Fig. 1.6 indicates an expected EMC of 0.2% for wood.

1.6 Specific Gravity and Density

The *specific gravity* (G) is the ratio of the oven-dry mass of a wood specimen to the mass of water displaced by the bulk specimen at a given moisture content. Since it is a ratio of masses, it is dimensionless. Specific gravity is numerically equal to oven-dry mass divided by moist volume. Expressed mathematically,

$$G = \frac{w_0}{V \varrho_w}, \tag{1.15}$$

where V = moist volume, ϱ_w = normal density of water = 1 g/cm³ = 62.4 lb/ft³.

It is apparent from Eq. (1.15) that an increase in the bound-water moisture content will swell the wood, causing a decrease in the specific gravity. At the fiber saturation point, the specific gravity has a minimum value, G_f, which remains constant at higher moisture contents.

The *density* of wood is the mass per unit volume at a given moisture content. Since any increase in moisture content will increase the mass of the wood at a greater rate than it increases its volume, an increase in moisture content will result in a higher density. This increase occurs at a greater rate above the FSP because swelling has ceased. Density may be expressed as

$$\varrho = \frac{w}{V}, \tag{1.16a}$$

where w = mass under moist conditions = $w_0(1 + 0.01 M)$.
Then,

$$\varrho = \frac{w_0(1 + 0.01 M)}{V}. \tag{1.16b}$$

A relationship between density and specific gravity may be derived from Eqs. (1.15) and (1.16b).

$$\varrho = G (1 + 0.01 M) \varrho_w. \tag{1.17}$$

It is clear from Eq. (1.17) that density and specific gravity are numerically equal under oven-dry conditions if expressed in cgs metric units. As moisture content increases, density becomes numerically greater than specific gravity. The relationship at various oven-dry specific gravities is illustrated in Fig. 1.7.

1.7 Specific Gravity of the Cell Wall and Porosity of Wood

The specific gravity of the oven-dry cell wall has been measured using displacement methods by many investigators with the values obtained being influenced by the

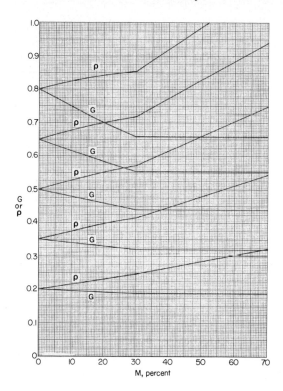

Fig. 1.7. Effect of moisture content on specific gravity and density, assuming constant lumen size and fiber saturation point of 30%. Density expressed in g/cm³

Table 1.9. Summary of measurements of cell-wall specific gravity and specific volume

Fluid	References	G_0'	v' cm³/g
Water	Stamm and Hansen (1937)	1.53	0.653
Water	Weatherwax and Tarkow (1968)	1.55	0.647
Water	Kellogg and Wangaard (1969)	1.50–1.53	0.653–0.667
Hexane by solvent exchange	Weatherwax and Tarkow (1968)	1.53	0.652
Toluene	Kellogg and Wangaard (1969)	1.45	0.690
Silicone oil and toluene	Weatherwax and Tarkow (1968)	1.47	0.682
Helium	Stamm and Hansen (1937)	1.46	0.685
Benzene	Stamm and Hansen (1937)	1.44	0.693
Mercury porosimetry	Stayton and Hart (1965)	1.44	0.693

properties of the displacing fluid. Typical of the results are those of Stamm and Hansen (1937), in which a value of 1.53 (specific volume 0.653 cm³/g) for water displacement, 1.46 (sp. v. 0.685 cm³/g) for helium, and 1.44 (sp. v. 0.693 cm³/g) for benzene as summarized in Table 1.9. They attributed the differences to two factors: (a) compression of bound water due to bonding forces at sorption sites in the case of water displacement and (b) the failure of nonpolar solvent molecules such as benzene to penetrate the microvoids in the cell wall. They assumed that the value obtained for helium was correct for wood substance. In order to account for the

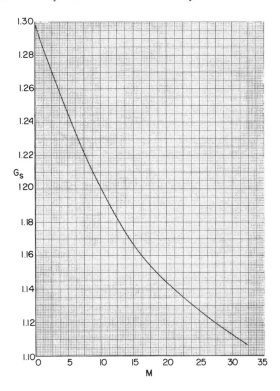

Fig. 1.8. Specific gravities of adsorbed water at different moisture contents according to Stamm and Seborg (1934). (Adapted from MacLean 1952)

different value for water displacement, an increased density for the bound water was assumed.

When comparing the specific volumes given above, there is a decrease of 0.685–0.653 = 0.032 cm³/g when going from helium to water. Assuming a FSP of 31%, the density of the bound water may be calculated as $0.31/(0.31-0.032) = 1.115$ g/cm³. The higher specific volume obtained with benzene displacement was explained by the failure of the latter to penetrate the microvoids. In this case the volume of the microvoids is 0.693–0.685 = 0.008 cm³/g. Therefore the percent microvoids is $0.008/0.685 \times 100\% = 1.2\%$. Stamm and Seborg (1934) measured the specific gravity of cell-wall substance at different moisture contents using benzene displacement, from which they calculated the apparent density of bound water over the hygroscopic range. Their values extended from 1.115 g/cm³, as calculated above for FSP, to approximately 1.3 g/cm³ under oven-dry conditions. The values are presented in Fig. 1.8. In later work with improved techniques Wilfong (1966) found no difference in the values obtained from helium and nonpolar solvents such as toluene.

Weatherwax and Tarkow (1968) attempted to explain the different values obtained from water and nonpolar solvents or helium on a basis of both of the factors cited by Stamm and Hansen (1937). They first measured the specific volume by displacement with the nonpolar silicone oil and toluene, obtaining equal values for each, and somewhat lower than Stamm and Hansen's value for benzene as revealed in Table 1.9. In an attempt to measure the volume of the microvoids only, they employed solvent displacement to replace water with ethanol, and then the ethanol with nonpolar hexane. This allowed the hexane to penetrate the microvoids, resulting in a specific volume which was 0.030 cm³/g smaller than that with silicone oil. This was assumed to be the volume of the microvoids. When the specific volume was measured by water displacement, the value of 0.647 cm³/g was 0.035 cm³/g smaller than with silicone oil. The decrease of 0.005 cm³/g observed with water displacement was assumed due to water compaction. Therefore it was concluded that approximately 15% of the difference in specific volumes was due to an increase in the apparent density of bound water and 85% to the presence of microvoids which could not be filled by the nonpolar solvent, except by using the technique of solvent exchange. The percent of microvoids would be 0.030/0.647 = 4.6% of the volume of the dry cell wall. Weatherwax and Tarkow concluded that the specific gravity of cell-wall substance of 0.647 cm³/g obtained by water displacement is relatively close to the true value, while that obtained by helium or toluene displacement is low.

Kellogg and Wangaard (1969) have presented an excellent literature review of the work in this field. They distinguish between wood substance, which does not include microvoids, and the cell wall, which includes them. The density of the latter on an oven-dry basis is approximately 1.45 g/cm³ as determined by toluene displacement, while their values for wood substance vary from 1.50 to 1.53 when corrected for some densification of water using the method of Weatherwax and Tarkow (1968). In the 18 species tested the calculated microvoid volume varied from 1.6% to 4.8%. Their values for wood substance are close to the densities measured by pycnometric displacement with water. Stayton and Hart (1965) measured cell-wall density by mercury porosimetry (see Chap. 4) where the displacing liquid was nonwetting mercury at a pressure of 3,000 lb/in², which would permit the penetration of voids larger than 0.03 μm. Their values were between 1.44 and 1.45 g/cm³, in close agreement with the those obtained by displacement by helium and nonpolar liquids. Since the mercury would not be expected to penetrate the microvoids, it is probable that the former fluids also did not penetrate them.

Kellogg et al. (1975) attempted to explain variations in the specific gravity of wood substance found in the 18 species investigated by Kellogg and Wangaard (1969) on a basis of differences in cell-wall composition. Stamm (1964) reported a density of 1.585 g/cm³ for alpha cellulose. Beall (1972) reported values of 1.46 and 1.67 g/cm³ for the hemicelluloses of hardwoods and softwoods, respectively, and 1.37 and 1.35 g/cm³ for the lignins of hardwoods and softwoods. A reasonable correlation was obtained when values were calculated from the constituents using the rule of mixtures. The presence of extractives also would be expected to influence cell-wall specific gravity. The specimens tested by Kellogg and Wangaard were all extracted using benzene-alcohol (2:1), alcohol, and distilled water to minimize this effect.

The specific gravity of oven-dry cell-wall or wood substance (G_0') is needed to calculate the *porosity*, or void volume fraction, and the swelling of wood. In view of the background of research in this field there is a practical necessity to decide on which value to use in the calculations. The value of 1.46 from the work of Stamm and Hansen (1937) may be used along with values of G_s taken from Fig. 1.8. The porosity of wood may then be calculated as

$$v_a = 1 - G\,(0.685 + 0.01\,M/G_s), \qquad (1.18)$$
$$M \leq M_f$$

where v_a = porosity or void volume fraction of wood, G = specific gravity of wood, $0.685 = 1/G'_o = 1/1.46$.

The term, 0.685 G in Eq. (1.18) is equivalent to G/G'_o, which is equal to the volume fraction of oven-dry cell wall (v_w), and the term 0.01 M G/G_s is equal to the volume fraction of bound water. The numeral 1 represents the total volume of the wood. Therefore v_a represents the void volume fraction or that not occupied by dry cell wall or moisture. It is clear that Eq. (1.18) is only valid up to M_f because any free water would have the normal density of 1 g/cm³.

The calculation of porosity is simplified if specific volumes based on water displacement are used because the specific gravity of bound water then becomes numerically equal to the normal density of water, 1 g/cm³, and the porosity calculation may be made up to the moisture content of fully saturated wood. Referring to Table 1.9, most of the values of G'_o are between 1.50 and 1.53 for water displacement. It is proposed that the value of 1.5 be selected for G'_o, corresponding to a specific volume of 0.667 cm³/g. Porosity may then be calculated as

$$v_a = 1 - G\ (0.667 + 0.01\ M). \tag{1.19}$$

(All moisture contents)
Porosity values calculated from Eq. (1.19) are in close agreement with those from Eq. (1.18), except at very low moisture contents. Equation (1.19) gives a porosity 0.6% higher at M = 6% and 0.4% lower at M = 30%.

Equation (1.19) may be used to calculate M_{max}, the moisture content of wood when all the cell walls and lumens are saturated by setting the porosity equal to zero.

$$M_{max} = 100/G - 66.7. \tag{1.20}$$

It is not known how the fraction of microvoids decreases with an increase in moisture content but it is probable that it decreases rapidly at low values of M. However, under oven-dry conditions the full fraction of microvoids will be present, and the porosity of the wood to a nonpolar gas or liquid will be lower than that to water. Therefore, under oven-dry conditions, it is preferable to use a specific volume of 0.685 cm³/g for nonpolar liquids.

$$v_a = 1 - 0.685\ G_o, \tag{1.21}$$

(Oven-dry porosity for sorption of nonpolar liquids)
and,

$$v_a = 1 - 0.667\ G_o, \tag{1.22}$$

(Oven-dry porosity for sorption of polar liquids).

Equations (1.18) to (1.22) may be used to calculate the maximum quantity of a liquid preservative which may be injected into a wood specimen of known specific gravity and moisture content. The porosity will determine the volume of liquid per unit volume of wood. This may be converted to a weight per unit volume basis by multiplying the porosity by the density of the liquid. It is sometimes useful to calculate preservative retention on a basis of the fraction of voids filled.

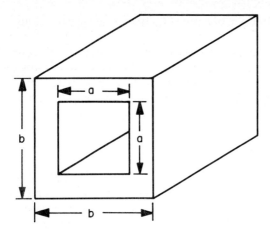

Fig. 1.9. Rectangular model for a cell wall

$$F_{VL} = \frac{w_L}{\varrho_L v_a V}, \qquad (1.23)$$

where F_{VL} = fraction of voids filled by liquids, w_L = mass of liquid, ϱ_L = density of liquid, V = volume of wood, v_a = porosity before impregnation.

Porosity may also be used to estimate the double cell-wall thickness of a softwood. If a rectangular model is assumed, as illustrated in Fig. 1.9, and all the cells are assumed of equal size and of square cross section, the porosity will be equal to the fractional cross-sectional area of the lumens.

$$v_a = a^2/b^2, \qquad (1.24)$$

where a = diameter of lumens, b = diameter of cells.

Then, by substitution into Eq. (1.19),

$$a^2/b^2 = 1 - G\,(0.667 + 0.01\,M). \qquad (1.25)$$

If a or b is known, the double cell wall thickness $(b-a)$ may be calculated.

1.8 Swelling and Shrinkage of the Cell Wall

It is assumed that M_f is 30%, that the specific gravity of wood substance is 1.5, and that the increase in volume due to swelling is equal to the volume of the bound water. Then, the increase in volume when M is increased from zero to M_f is $0.3\,(1.5) = 0.45\,\text{cm}^3$ per cm^3 of cell wall. Therefore the maximum swelling of the cell wall is 45%. In general it may be written that

$$S' = M\,G'_0 \qquad (1.26)$$

where S' = percent swelling of the cell wall from oven-dry to M.

The maximum percent swelling to M_f is then

$$S'_f = M_f\,G'_0. \qquad (1.27)$$

The specific gravity of the cell wall at any moisture content within the hygroscopic range may be calculated from the oven-dry value as

$$G' = \frac{G'_0}{1 + 0.01 \, M \, G'_0}. \tag{1.28}$$

The specific gravity of the cell wall at M_f is then

$$G'_f = \frac{G'_0}{1 + 0.01 \, M_f G'_0}. \tag{1.29}$$

The value of G'_f will then be equal to $1.5/[1+0.3(1.5)] = 1.034$, assuming $M_f = 30\%$. The density at M_f (ϱ'_f) will then be equal to $1.034\,(1.3) = 1.346$ g/cm³ from Eq. (1.17). Therefore, the density of the cell wall decreases with an increase in moisture content due to the lower density of water.

The shrinkage of the cell wall from M_f may be calculated as

$$s' = (M_f - M) \, G'_f, \tag{1.30}$$

where s' = percent shrinkage of cell wall from M_f to M.

The maximum percent shrinkage from M_f to oven-dry is

$$s'_f = M_f \, G'_f. \tag{1.31}$$

If M_f is assumed to be 30%, $s'_f = 30\,(1.034) = 31.0\%$. This is a lower percent than the 45% maximum swelling because it is based upon a larger initial volume.

1.9 Swelling and Shrinkage of Wood

Two simplifying assumptions are made in the analysis of the shrinkage and swelling of wood. First, the lumens of the cells are assumed to be of constant size. Beiser (1933) measured a negligible change in the areas of the lumens of the wood of *Picea* sp. and *Betula* sp. with changes in the bound-water content of the cell walls. This assumption does not hold for all woods because, in some, the lumens decrease in size while, in others, they increase with an increase in moisture content. However, in most cases the assumption is a reasonable one. Based on constant lumen area, the change in the volume of a wood specimen is equal to the volume of the bound water added to or removed from the cell wall. Secondly, the fiber saturation point is assumed to be constant for all woods. In this second assumption, a higher or lower M_f would change the expected swelling or shrinkage proportionally.

Assume an oven-dry cell illustrated schematically in cross section in Fig. 1.10. When bound water is added the outer dimensions are extended to the outside of the cross-hatched area. It may be stated that

$$V_1 = V_0 + \Delta V, \tag{1.32}$$

where V_0 = oven-dry volume, V_1 = volume at M_1, below FSP, ΔV = increase in volume due to swelling.

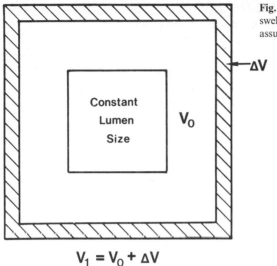

Fig. 1.10. Schematic illustration of the swelling of the cross section of a wood cell assuming constant lumen size

It is assumed that the increase in volume is entirely due to bound water entering the cell wall. Then

$$\Delta V = V_0 G_0 (0.01 M_1), \tag{1.33}$$

where G_0 = oven-dry specific gravity.

Then Eq. (1.33) may be substituted into Eq. (1.32)

$$V_1 = V_0 (1 + 0.01 M_1 G_0). \tag{1.34}$$

The maximum swollen volume of the wood may be calculated from Eq (1.34) as

$$V_f = V_0 (1 + 0.01 M_f G_0). \tag{1.35}$$

Assuming $M_f = 30\%$

$$V_f = V_0 (1 + 0.30 G_0) \tag{1.36}$$

It is clear from Eq. (1.36) that the maximum volumetric swelling of wood is directly proportional to the oven-dry specific gravity, assuming constant fiber saturation point and lumen size.

Similarly, it may be shown that the fractional volumetric shrinkage is directly proportional to the specific gravity at the fiber saturation point. Then the oven-dry volume may be calculated from the swollen volume as

$$V_0 = V_f (1 - 0.30 G_f). \tag{1.37}$$

From Eq. (1.36) it follows that

$$S_f = 30 G_0, \tag{1.38}$$

where S_f = percent volumetric swelling of wood from oven-dry to M_f. And from Eq. (1.37),

$$s_f = 30 G_f, \tag{1.39}$$

where s_f = percent volumetric shrinkage from M_f to oven-dry.

Since G_f for a given wood is less than G_0, the swelling percentage is less than that of shrinkage because it is based upon a smaller initial volume.

Equations (1.38) and (1.39) predict that swelling and shrinkage are greater in the denser woods and are directly proportional to specific gravity, or the quantity of cell-wall substance present. This is based upon the assumption of constant M_f. Kellogg and Wangaard (1969) observed that woods of lower specific gravity tend to have higher fiber saturation points, decreasing the expected difference in swelling between low- and high-density woods.

Equations (1.36) and (1.37) are the basis for writing a relationship between G_0 and G_f for a given wood specimen. By definition, specific gravity is inversely proportional to volume. Expressed mathematically,

$$G_0 V_0 = G_1 V_1 = G_2 V_2 = G_f V_f, \tag{1.40}$$

where the respective values of G and V are for the same wood specimen at moisture contents M_0, M_1, M_2, and M_f.

Then

$$G_0 = \frac{G_f}{1 - 0.30 G_f} \tag{1.41}$$

and

$$G_f = \frac{G_0}{1 + 0.30 G_0}. \tag{1.42}$$

Similarly, for converting a specific gravity from moisture content M_1 to M_2,

$$G_2 = \frac{G_1}{1 + 0.01 G_1 (M_2 - M_1)}. \tag{1.43}$$

M_1 and $M_2 \leq M_f$.

The anisotropy of wood results in different shrinkage in the various directions. The longitudinal shrinkage of normal wood is 0.1% to 0.3%, although it may be much higher in juvenile, compression, or tension woods. For normal wood the longitudinal shrinkage is negligible. The shrinkage in the tangential direction (T) is usually double that in the radial direction (R) in most woods. Therefore an assumption is made that the T/R ratio is 2.0 and that two thirds of the volumetric shrinkage or swelling is tangential and one third is radial. The decreased shrinkage in the radial direction has been attributed by Panshin and de Zeeuw (1980) to (a) the restriction of radial shrinkage due to the presence of the ray cells and (b) the presence of bands of low-density earlywood and high-density latewood. The effect of these two segments is additive in the radial direction but, in the tangential direction, the denser and stronger latewood controls the shrinkage and forces the earlywood to shrink by essentially the same amount.

Equations (1.38) and (1.39) may be modified for the calculation of theoretical tangential and radial shrinkage and swelling percentages.

$$S_{fT} = 20 G_0, \tag{1.44}$$

where S_{fT} = percent tangential swelling from oven-dry to M_f

$$S_{fR} = 10 G_0, \tag{1.45}$$

where S_{fR} = percent radial swelling from oven-dry to M_f

$$s_{fT} = 20\, G_f, \tag{1.46}$$

where s_{fT} = percent tangential shrinkage from M_f to oven-dry

$$s_{fR} = 10\, G_f, \tag{1.47}$$

where s_{fR} = percent radial shrinkage from M_f to oven-dry.

It is approximately correct that

$$S_f \approx S_{fT} + S_{fR} \tag{1.48}$$

and

$$s_f \approx s_{fT} + s_{fR} \tag{1.49}$$

It is assumed that swelling and shrinkage are linear with moisture content over the hygroscopic range. Then the swelling or shrinkage between any two moisture contents may be calculated as

$$S = G_1\,(M_2 - M_1)/30 \tag{1.50}$$

$$s = G_2\,(M_2 - M_1)/30, \tag{1.51}$$

where S = swelling percent from M_1 to M_2, s = shrinking percent from M_2 to M_1, $M_2 > M_1$, and M_1 and $M_2 \leq M_f$.

Chapter 2
Wood Structure and Chemical Composition

2.1 Introduction

Wood is extremely nonhomogeneous, and its structural and chemical variability is reflected in wide ranges in its physical properties such as permeability, capillary behavior, thermal conductivity, and the diffusion of bound water. The greater uniformity is found among the woods of conifers or *softwoods*. Those of dicotyledonous angiosperms or *hardwoods*, on the other hand, have extreme structural differences and are therefore much easier to distinguish from one another visually. Hardwood species are also much more numerous because they occur in tropical as well as temperate regions, while softwoods are found primarily in the temperate zones. Because of greater structural variation the hardwoods have a greater range in permeability and capillary behavior. In both types *earlywood* is generally of much lower density than *latewood*. The earlywood of dried softwoods is usually less permeable than the latewood while, in hardwoods, the reverse is commonly true owing to the larger earlywood vessels. There can also be significant differences between stands of trees of the same species, within a tree, between boards, and within a board. *Heartwood* is nearly always much less permeable than *sapwood* due to pit aspiration, incrustation, and vessel tyloses in the former.

Compression wood in conifers and *tension wood* in hardwoods have physical properties and chemical compositions vastly different from normal wood. Tracheid and ray cells in softwoods have dissimilar chemical compositions and the same applies to the fibers, vessels, and ray cells of hardwoods. On the ultrastructural level the middle lamella (intercellular substance), the primary wall, and the secondary wall have significantly different chemical compositions. Within the secondary wall there are also differences in composition between the S_1, S_2, and S_3 layers.

Wood consists of three major components: *cellulose* is the skeleton, *hemicellulose* the matrix, and *lignin* the encrusting substance which binds the cells together and gives rigidity to the cell wall. Additionally, heartwood in particular contains many low-molecular-weight organic compounds known as *extractives* or extraneous substances. These are commonly located in the lumens but may also occur in the cell wall as in *Sequoia sempervirens*. A few, such as tannins, are polymeric. The content of extractives can vary from near zero to 25% by weight in temperate-zone species and much higher in some tropical hardwoods. Inorganic constituents or *ash* seldom exceed 0.1% to 0.5%.

Two recent books provide excellent background material, namely Panshin and de Zeeuw (1980) for wood structure and Sjöstrom (1981) for wood chemistry.

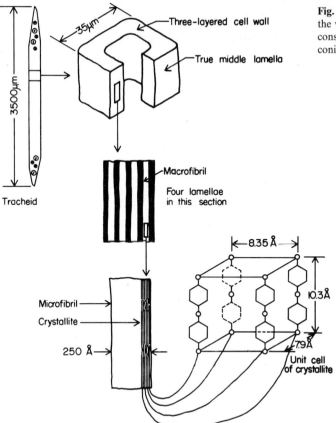

Fig. 2.1. Relationship between the various elements constituting the cell wall in coniferous wood. (Siau 1971)

2.2 The Cell Wall

The relationship between the various elements making up the cell wall in a longitudinal conifer tracheid is outlined in Fig. 2.1. The microfibrils (see Sect. 2.10.1) are aggregated into longer macrofibrils, organized into lamellae within the different layers of the wall. The organization of a softwood tracheid or hardwood fiber is shown schematically in Fig. 2.2. According to the interrupted lamella model proposed by Kerr and Goring (1975), cellulose microfibrils, coated with hemicelluloses, are embedded in matrix of hemicelluloses and lignin, as illustrated in Fig. 2.3.

The secondary wall in tracheids, fibers, and vessels of normal wood consists of three layers which are referred to as the S_1, S_2, and S_3 layers. Careful measurements of the thickness in radial and tangential directions of these layers in earlywood and latewood of various conifer species have been reported by Saiki (1970). The greater wall thickness of latewood in comparison with earlywood is largely a result of an increase in the S_2 layer on transition from earlywood to latewood.

The middle lamella-primary wall region (M+P) varies in width from 0.1 to 0.4 μm along the wall. At the cell corners the intercellular space can be 3 to 4 μm.

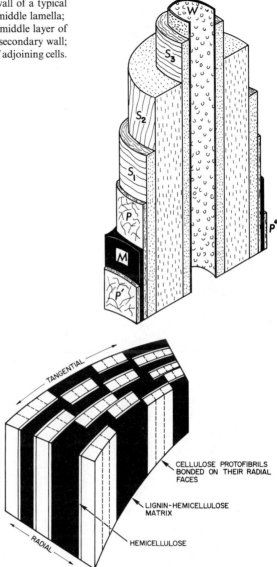

Fig. 2.2. Diagrammatic view of the cell wall of a typical coniferous tracheid. *P* primary wall; *M* middle lamella; S_1 outer layer of the secondary wall; S_2 middle layer of the secondary wall; S_3 inner layer of the secondary wall; *W* warty layer; *P'* and *P''* primary walls of adjoining cells. (Ward et al. 1969)

Fig. 2.3. The interrupted lamella model, showing the arrangement of cellulose, hemicellulose, and lignin in the wood cell wall. (Kerr and Goring 1975)

The primary wall itself is only 0.1 to 0.2 μm thick. The orientation of the cellulose microfibrils is essentially random in this wall. The S_1 layer is 0.2 to 0.5 μm wide in earlywood but can reach a width of 1.0 μm in latewood. Its microfibrils are oriented at an average angle of 60° to 80° to the fiber axis.

The dominating S_2 layer has a thickness that varies from 1.0 to 2.0 μm in earlywood and from 3 to 8 μm in latewood. In the former wood the microfibril angle is about 10°, while reaching 30° in latter. When water enters between the cellulose chains in the S_2 layer it forces the chains apart, causing transverse (radial and tangential) swelling, while any change in the longitudinal direction is minor. The swell-

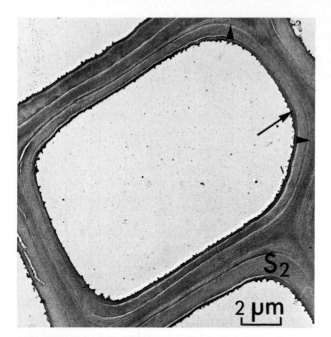

Fig. 2.4. Normal wood tracheids in the earlywood region of *Tsuga canadensis*, showing the large size of the cells and their thin walls. *Short arrows* indicate the S_1 layer. *Long arrow* indicates the S_3 and warty layers. Transverse section. TEM

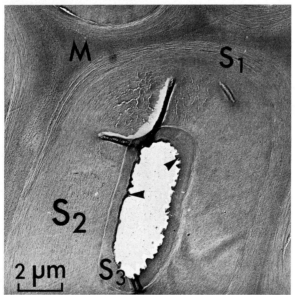

Fig. 2.5. Normal wood tracheids in the latewood region of *Tsuga canadensis*, showing a tracheid with a very thick wall. *M* middle lamella; *Short arrows* indicate warts. Transverse section. TEM

ing and shrinking of wood in this direction increases rapidly with increasing microfibril angle in S_2 (Barber and Meylan 1964; Harris and Meylan 1965; Barber 1968; Meylan 1968, 1972) although other factors, such as the nature of the cell wall constituents, must also be taken into account (Cave 1972a, b, 1976, 1978a, b).

The S_3 layer is only 0.1 to 0.2 µm thick in both earlywood and latewood. Its microfibrils are almost transverse to the fiber axis. This layer is lined on the lumen

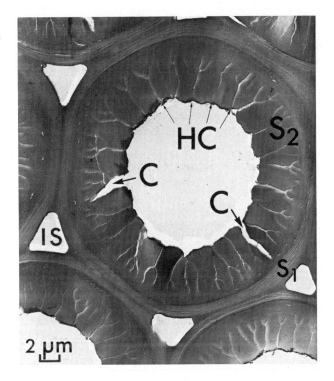

Fig. 2.6. Compression wood tracheids in *Larix laricina* with a rounded outline and thick wall. *IS* intercellular space; *HC* helical cavities; *C* drying checks. Transverse section. TEM

side in most conifers by a so-called warty layer of unknown origin and chemical composition. Figures 2.4 and 2.5 show the appearance in transverse sections of the M+P, S_1, S_2, and S_3 layers in earlywood and latewood tracheids of *Tsuga canadensis*.

Because the S_2 constitutes 60% to 80% of the total secondary wall, this layer has the greatest influence on the shrinkage and swelling behavior of wood with S_1 and S_3 acting as constricting elements. When S_2 is more equal to S_1 in width, as it often is, for example, in tension wood fibers, the influence of the S_1 layer cannot be neglected. The warty layer has considerable influence on the diffusion of water through the cell wall.

The S_1 layer in the typically rounded compression wood tracheids is considerably thicker than in corresponding normal wood, usually 0.3 to 0.5 µm. It is often very massive at the cell corners, where it can reach a thickness of 2 to 5 µm. The microfibril angle in this layer is 80° to 90°. Except for a few first-formed and last-formed cell rows, compression wood tracheids vary less in width than those in normal wood. A thickness of 3 to 5 µm is most common in S_2. The microfibrils are oriented at an angle of 30° to 50° to the fiber axis in this layer, a fact that is largely, albeit not entirely, responsible for the exceptionally high longitudinal shrinkage of compression wood (Harris and Meylan 1965). The most characteristic feature of this layer is the presence of helical cavities and ribs which are oriented in the same direction as the microfibrils. The appearance of the wall layers in a compression wood tracheid is seen in Fig. 2.6, which also shows the occurrence

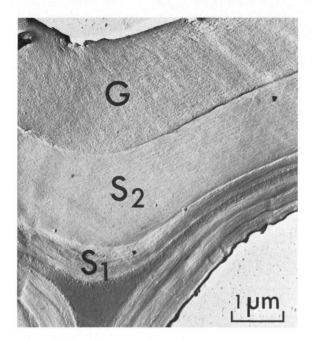

Fig. 2.7. Tension wood fibers in *Fagus grandifolia*. *G* gelatinous layer, Transverse section. TEM

of the intercellular spaces, another characteristic feature of compression wood. There is no S_3 layer in fully developed compression wood tracheids.

The gelatinous layer or G-layer in tension wood most often replaces the S_3 layer in these fibers. As shown in Fig. 2.7, it can rival or surpass S_2 in thickness. It consists of highly crystalline cellulose microfibrils, oriented parallel with the fiber axis. Because of this circumstance and the fact that this layer is only loosely attached to S_2, the G-layer has no influence on the longitudinal shrinkage or swelling of tension wood (Norberg and Meier 1966).

2.3 Structure of Softwoods

A typical softwood structure is depicted in Fig. 2.8, representing the wood of a hard pine, *Pinus taeda*. The prosenchyma or fluid-conducting tissue consists of the longitudinal and ray tracheids. The parenchyma which functions in the living tree as storage tissue for reserve food material and radial transport includes the longitudinal parenchyma, the epithelial cells surrounding resin canals, and the ray pa-

---

Fig. 2.8. Gross structure of a typical southern pine softwood. *Transverse view.* *1–1 a* ray; *B* dentate ray tracheid; *2* resin canal; *C* thin-walled longitudinal parenchyma; *E* epithelial cells; *3–3 a* earlywood tracheids; *F* radial bordered pit pair cut through torus and pit apertures; *G* pit pair cut below pit apertures; *H* tangential pit pair; *4–4 a* latewood.
Radial view. *5–5 a* sectioned fusiform ray; *J* dentate ray tracheid; *K* thin-walled parenchyma; *L* epithelial cells; *M* unsectioned ray tracheid; *N* thick-walled parenchyma; *O* latewood radial pit (inner aperture); *O'* earlywood radial pit (inner aperture); *P* tangential bordered pit; *Q* callitroid-like thickenings;

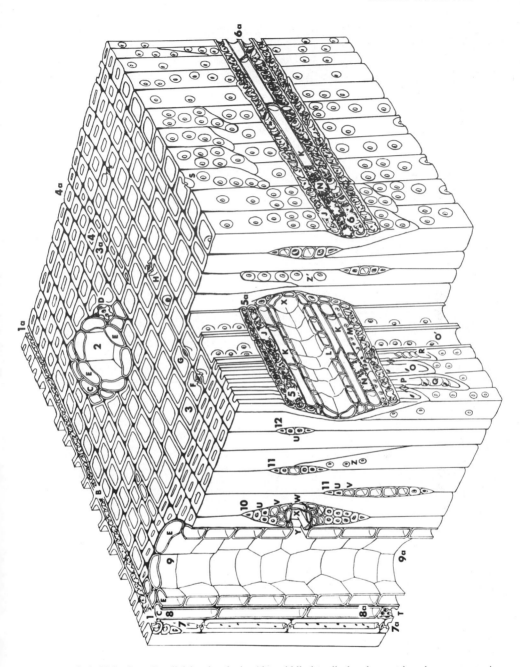

R spiral thickening; S radial bordered pits (the middle lamella has been stripped away, removing crassulae and tori); $6-6a$ sectioned uniseriate heterogeneous ray.
Tangential view. $7-7a$ strand tracheids; $8-8a$ longitudinal parenchyma (thin-walled); T thick-walled parenchyma; $9-9a$ longitudinal resin canal; 10 fusiform ray; U ray tracheids; V ray parenchyma; W horizontal epithelial cells; X horizontal resin canal; Y opening between horizontal and vertical resin canals; 11 uniseriate heterogeneous rays; 12 uniseriate homogeneous ray; Z small tangential pits in latewood; Z' large tangential pits in earlywood. (Howard and Manwiller 1969)

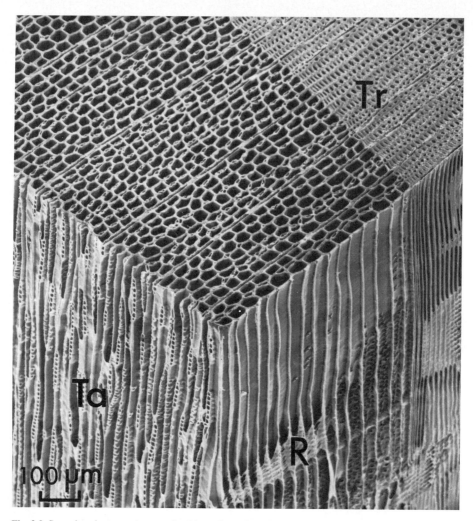

Fig. 2.9. Scanning electron micrograph of the softwood *Larix occidentalis*, showing an abrupt transition between earlywood and latewood. The thin-walled earlywood tracheids are seen in the *foreground* on the transverse section *(Tr)* and the thick-walled latewood tracheids in the *upper right*. The tangential cell diameters are approximately equal, but the latewood cells have a much smaller radial diameter than those in the earlywood. Abundant intertracheid pitting is visible on the radial surface *(R)*. Uniseriate rays are visible on the transverse surface and in transverse section on the tangential surface *(Ta)*. (Courtesy of the N.C. Brown Center for Ultrastructure Studies)

renchyma. Longitudinal and horizontal resin canals are also present in *Pinus*, *Picea*, *Larix*, and *Pseudotsuga*. These are not considered as cells since they are intercellular spaces. A scanning electron micrograph of *Larix occidentalis*, also with an abrupt transition between earlywood and latewood, is shown in Fig. 2.9.

Longitudinal tracheids make up the bulk of the structure of gymnospermous woods. Earlywood and latewood tracheids are illustrated in Fig. 2.10. These are long, imperforate, narrow cells with which have ends that are tapered along the

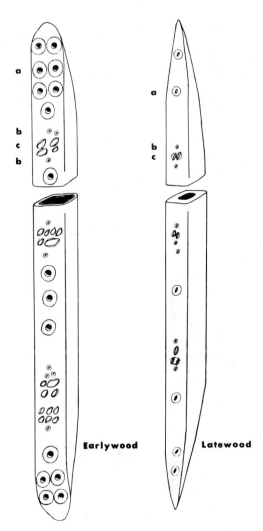

Fig. 2.10. Radial surfaces of earlywood (*left*) and latewood (*right*) tracheids. *a* intertracheid bordered pits; *b* bordered pits to ray tracheids; *c* pinoid pits to ray parenchyma. (Howard and Manwiller 1969)

radial surfaces for a considerable portion of their lengths where they are in contact with other tracheids. Most of the intertracheid bordered pit pairs are along these tapered portions of the radial surfaces. There are smaller and fewer pits in latewood tracheids than in earlywood and generally fewer pits of smaller size on the tangential surfaces of all longitudinal tracheids; thus, most of the fluid flow between tracheids takes place in the tangential direction if flow through the rays is not considered. The number of pits per tracheid varies from 50 to 300 in earlywood, with fewer in latewood (Stamm 1964).

The cell-wall cross sections of earlywood and latewood may be compared in the electron micrographs of Figs. 2.4 and 2.5. It is clear from these that the lumens are much larger and the cell-wall layers much thinner in earlywood. The radial diameter of tracheids is greatest in the earlywood, with the latewood having thicker-

Table 2.1. Summary of dimensions of structural elements in normal softwoods

Structural element	Dimensions, μm
Tracheid length	3,500
Tracheid diameter	35
Tracheid double cell wall thickness in latewood	10
Tracheid lumen diameter	20–30
Thickness	
True middle lamella in latewood	0.1–4.0
Primary wall	0.1–0.2
S_1 layer in latewood	1
S_2 layer in latewood	3–8
S_3 layer	0.1–0.2
Overall diameter of pit chambers of bordered pits	6–30
Effective diameter of pit openings	0.02–4.00

walled and narrow cells. The tangential diameter may vary from 15 to 80 μm according to species, and the length may be in the range of 1,200 to 7,500 μm (1.2 to 7.5 mm). Average values of diameter and length may be taken as 35 μm and 3,500 μm, with a length-to-diameter ratio of approximately 100. A dimensional summary of softwood structural elements is given in Table 2.1. The above considerations give rise to the Comstock model described in Chap. 3.10. There are bordered pit pairs leading from longitudinal tracheids to the ray tracheids and half-bordered pit pairs permitting flow to adjacent ray parenchyma cells. Most of the radial flow of fluids is probably through the ray cells and between the longitudinal tracheids and rays.

The approximate volumetric composition of the wood of *Pinus strobus* (Panshin and de Zeeuw 1980) is

Longitudinal tracheids 93%
Longitudinal resin canals 1%
Wood rays 6%

This is a typical softwood composition. The same authors give a range of ray volumes from 3.4% to 11.7% for softwood species with an average of 7%, contrasted with an average of 17% for hardwoods.

Since the rays and resin canals form but a small fraction of the volume, their contributions to the overall flow may be of secondary importance. If these flow paths are neglected, a very simple flow model for softwoods results, in which fluids flow from tracheid to tracheid through the bordered pit pairs. This is in general agreement with microscopic observations (Wardrop and Davies 1961; Behr et al. 1969; Bailey and Preston 1969). There is some evidence, however, that rays may be an important flow path for preservatives into long, thin specimens. If there were no flow through the rays, the tangential permeability would exceed the radial because most of the pits are on the radial surfaces of the tracheids. However, Banks (1970) has found the radial permeability of *Pinus sylvestris* wood to be greater than the tangential permeability, and has concluded that this is due to flow through the rays and that it explains the superior treatability of this wood when compared with *Picea abies*.

The longitudinal permeability of softwoods is much greater than the tangential permeability, due to the fact that there are fewer pitted cross walls to traverse per unit length in the longitudinal direction.

The relatively high ratio of tangential to radial shrinkage of wood (2:1) can be explained by considering the structure. The thicker latewood tracheids have a higher proportion of material in the S_2 layer where the microfibrils are oriented longitudinally, resulting in a higher fractional transverse shrinkage in latewood than in earlywood tracheids. The latewood tracheids predominate in controlling tangential shrinkage since they contain more cell-wall substance, thus forcing the earlywood tracheids to comply with their dimensional changes. In the radial direction, the earlywood and latewood shrinkages are additive so the smaller shrinkage of earlywood tissue results in a reduction in the over-all fractional radial shrinkage. Radial shrinkage is also constrained by the presence of ray cells.

2.4 Types of Pit Pairs

A pit pair consists of complimentary gaps or recesses in the secondary walls of two adjacent cells, together with a pit membrane that consists of a middle lamella sandwiched between two primary walls. There are three basic types of pit pairs which are found in both softwoods and hardwoods, illustrated in Fig. 2.11. The simple pit pair (a) is usually located between two parenchyma cells. An example of such a pit pair between two ray parenchyma cells of *Larix laricina* wood is shown in Fig. 2.12. A half-bordered pit pair (c) is located between a parenchyma and a prosenchyma cell and the bordered portion facing the prosenchyma cell. This is illustrated in Fig. 2.13 by a pit pair between a ray parenchyma and ray tracheid cell of *Pinus rigida* wood. A bordered pit pair (b) is situated between two prosenchyma cells such as ray tracheids or longitudinal tracheids, illustrated in Figs 2.14 to 2.20 and 2.25 to 2.27.

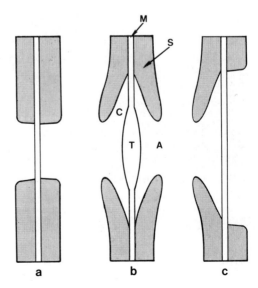

Fig. 2.11. Three basic types of pit pairs. **a** simple pit pair; **b** bordered pit pair; **c** half-bordered pit pair. *A* aperture; *C* chamber; *M* middle lamella-primary wall; *S* secondary wall; *T* torus

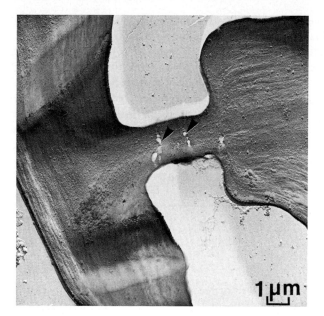

Fig. 2.12. Simple pit pair between ray parenchyma cells in *Larix laricina*. Openings (*arrows*) are former plasmodesmata. Thickness of membrane: 1.8 µm, diameter of openings in membrane: 0.1 to 0.3 µm. TEM by L.P. Mann. (Siau 1971)

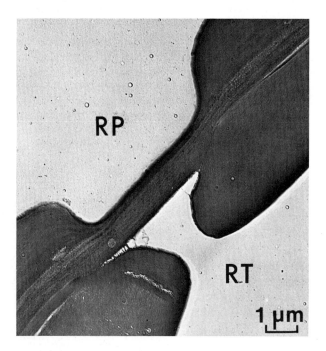

Fig. 2.13. Half-bordered pit pair between a ray tracheid *(RT)* and a ray parenchyma *(RP)* cell in latewood of *Pinus rigida*. Thickness of pit membrane: 1.2 µm. Transverse section. TEM. (Côté and Day 1969)

Fig. 2.14. Aspirated earlywood bordered pit pair between tracheids in heartwood of *Pinus echinata*. The torus *(T)* is held thightly against the pit border *(PB)*, and its surface is dished, probably by capillary forces acting during drying. Approximate dimensions: pit aperture *(PA)* diameter: 4 µm, inside diameter of pit chamber: 17 µm, double cell wall thickness, 3 µm. Tangential section. TEM by A.C. Day. (Siau 1971)

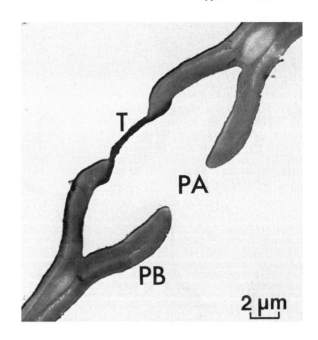

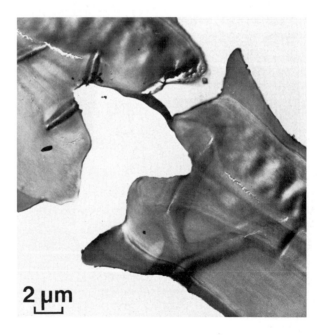

Fig. 2.15. Aspirated bordered pit pairs between tracheids in heartwood of *Pinus elliottii*. The torus is drawn and held tightly to the pit border and is dished. It is smaller and thicker than the earlywood torus in Fig. 2.14. Aperture diameter: 2 to 3 µm, inside diameter of chamber: 10 µm, double cell wall thickness: 12 µm. Tangential section. TEM by A.C. Day. (Siau 1971)

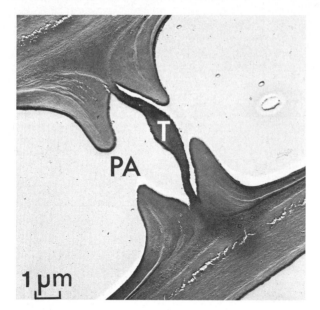

Fig. 2.16. Earlywood bordered pit pair between two ray tracheids in heartwood of *Pinus glabra*. The relatively thick pit membrane (0.3 μm) and its torus *(T)* have been displaced from the central, unaspirated position, probably by capillary forces during drying. Aperture diameter: 1.7 μm, inside diameter of chamber: 5.3 μm. *PA* pit aperture. Tangential section. TEM by A.C. Day. (Siau 1971)

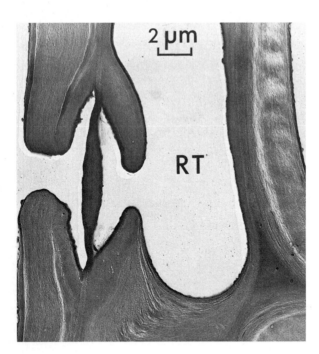

Fig. 2.17. Earlywood bordered pit pair in the heartwood region of *Pinus pungens* between a ray tracheid *(RT)* to the *right* and an axial tracheid to the *left*. Transverse section. TEM. (Côté and Day 1969)

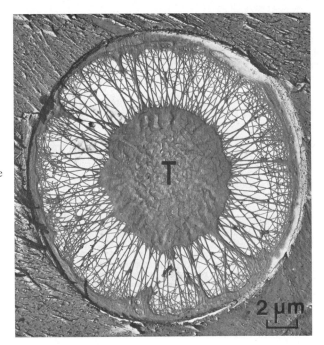

Fig. 2.18. Unaspirated bordered pit in a sapwood tracheid of *Tsuga canadensis*, dried by solvent exchange from acetone to prevent aspiration. The back border was torn away during replication, revealing the fibrillar structure of the margo. Inside diameter of the chamber: 18 µm, diameter of the torus (T): 9 µm, openings in the margo: up to 1 µm in the circumferential direction and up to 2 µm in the radial direction. TEM replica. (Comstock and Côté 1968)

2.5 Softwood Pitting

Of the three types of pit pairs, the bordered pit pair has the most significant influence on flow properties, because nearly all of the softwood tissue consists of prosenchyma cells. Pit pairs are depicted in cross section in Figs. 2.14 to 2.17, while Figs. 2.18 to 2.20 illustrate the surface of the membrane. The overall diameter of the pit chambers of conferous bordered pits has an approximate range of 6 to 30 µm, earlywood pits being larger than latewood pits. The diameter of the torus, when present, is one-third to one-half the overall diameter of the chamber, and that of the aperture is approximately one-half the diameter of the torus. The torus is the thickened center portion of the pit membrane, consisting of primary wall material. It is typical of species of the Pinaceae family. Usually there are no apparent openings in the torus, but occasionally small holes are observed in certain species. Generally the torus is impermeable to fluids, as shown by the established correlation between pit aspiration and low permeability. The membrane surrounding the torus is called the margo, and it consists of strands of cellulose microfibrils radiating from the torus to the periphery of the pit chamber as illustrated in Fig. 2.18. The openings between the microfibrils permit the passage of fluids and small particles through the pit membrane. Microscopic measurements and indirect methods such as calculation from gas flow (Chaps. 3.8 to 3.10) and from capillary pressures (Chaps. 4.3 and 4.4) indicate that softwood pit openings have effective diameters between 0.02 and 8 µm. The larger values usually result from flow measurements and may reflect the size of the aperture rather than the openings in the pit membrane, particularly in latewood where the aperture resembles a capillary rather

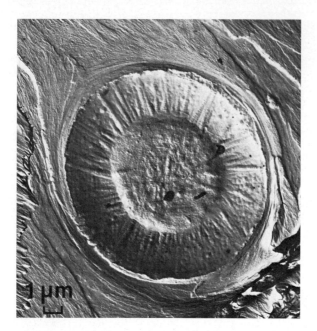

Fig. 2.19. Earlywood bordered pit pair in the heartwood region of *Pinus rigida*. The pit membrane is heavily encrusted and aspirated, and the torus is dished into the aperture. TEM replica by A.C. Day. (Siau 1971)

than an orifice (Figs. 2.14 and 2.15). The flow resistance of the aperture under these conditions could exceed that of the combined pit-membrane openings; therefore calculations from the flow measurements would tend to favor the aperture diameter. Sebastian et al. (1965) obtained calculated diameters of 1.4 to 5.0 µm with an average of 2.6 µm for pit-membrane openings in the sapwood of *Picea glauca* from flow measurements, while microscopic examination yielded an average distance between the microfibrillar strands of 1.3 µm. Therefore the values from flow measurements were approximately twice those measured with the microscope in sapwood, and they were four times as great in heartwood. Figure 2.18 reveals maximum interfibrillar openings 1 µm wide and 2 µm long in this specimen of *Tsuga canadensis* wood (Comstock and Côté 1968). Petty (1970) measured openings 0.25 µm wide and 1 µm long, with an equivalent diameter of 0.4 µm, from electron micrographs of *Picea sitchensis* wood. It may accordingly be concluded that the approximate range of pit opening diameters in softwoods probably extends from 0.02 µm to 4 µm (20 to 4,000 nm) with a logarithmic mean of 0.3 µm.

The thickness of the pit membrane is an important dimension affecting flow resistance and the interpretation of flow data. Sebastian et al. (1965) assumed a value of 0.15 µm for the membrane in *Picea glauca* wood and Petty (1970) estimated a value of 0.1 µm in his calculations of pit-membrane sizes in *Picea sitchensis* wood. The pit membrane illustrated in Fig. 2.16 is approximately 0.3 µm thick and that in Fig. 2.17 is 0.5 µm. Although both of these are from earlywood, they are also heartwood and may have been thickened by incrustations. The torus thickness in the earlywood pit pair illustrated in Fig. 2.14 is 0.3 µm and that of the latewood pit pair in Fig. 2.15 is 0.4 µm. Although these pits are aspirated and the membranes are not visible, their thicknesses would be much less than the above values. An ap-

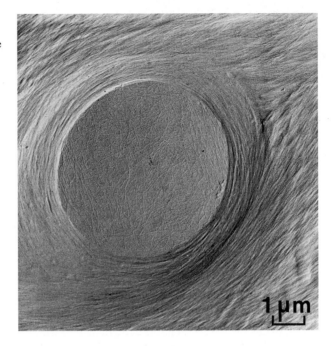

Fig. 2.20. Structure of the torus surface as observed through the aperture from the lumen side of a tracheid in *Pinus elliottii*. Diameter of aperture: 6 µm. TEM replica by A.C. Day. (Siau 1971)

proximate range of margo thicknesses may be assumed to extend from 0.1 to 0.5 µm with a logarithmic mean of 0.2 µm.

The membranes of simple and half-bordered pit pairs are apparently much thicker than those in bordered pit pairs. The simple pit pair in Fig. 2.12 has a membrane 1.8 µm thick, and that of the half-bordered pit pair in Fig. 2.13 is 1.2 µm thick. Despite relative thickness and lack of openings in these membranes, there is evidence of their effectiveness as flow paths as discussed in the next section.

Earlywood and latewood bordered pit pairs are compared in Figs. 2.14 and 2.15; the latter shows a smaller overall diameter with a thicker torus and membrane resulting in a more rigid structure which is much better able to resist the capillary forces causing aspiration as discussed in Chap. 4.6. The latewood apertures resemble capillary tubes while those in earlywood are like orifices. This variation could produce significant differences in flow properties.

Côté (1963) described changes in pit membranes of coniferous woods, particularly during heartwood formation, which account for decreased permeability. The first is *pit aspiration*, shown in Figs. 2.14, 2.15, and 2.19, in which the torus is dished and is held rigidly against the overarching border, blocking the aperture and preventing the penetration of fluids. The torus is probably held in this position by hydrogen bonds between adjacent cellulose chains. The second condition is *incrustation* with extractives, as illustrated in Fig. 2.19.

The cell-wall structure surrounding a pit aperture viewed from the interior of the lumen is shown in Fig. 2.20. The randomly oriented microfibrillar structure of the torus is evident, along with portions of the secondary wall layers in the overarching border surrounding the aperture.

2.6 Microscopic Studies of Flow in Softwoods

Microscopic observations have made it possible to elucidate the flow paths in softwoods resulting from treatment with both oily or nonpolar liquids and aqueous solutions. Behr et al. (1969) treated four softwood species with creosote and pentachlorophenol dissolved in aromatic gas oil followed by examination under an optical microscope. They found many longitudinal tracheid lumens and bordered pit pairs filled with both preservatives in all the species, with more penetration in the latewood than in the earlywood in most cases. Oil was also found in ray tracheids and ray parenchyma cells and in earlywood longitudinal tracheids near the rays in many instances. The passage of oil between ray parenchyma cells and from ray parenchyma to longitudinal tracheids was evidence of flow through the membranes of simple and half-bordered pit pairs. Rays were found to be a more important flow path in the softwoods than in the hardwoods which were similarly investigated. Vertical and horizontal resin canals and epithelial cells were generally filled. The decreased penetration of heartwood compared with sapwood was attributed to the higher fraction of aspirated pits in the former.

Wardrop and Davies (1961) treated dried specimens of *Pinus radiata* with aqueous salt solutions followed by suitable precipitating agents with the flow alternatively limited to the longitudinal and radial directions. In the former case, flow was detected between longitudinal tracheids through bordered pit pairs and laterally through ray cells. The vertical resin canals also made a significant contribution. When the flow was directed radially, the ray parenchyma cells were apparently more easily penetrated than the ray tracheids, contrary to most comparisons of flow through these cell types. Liquid also flowed from the rays to the longitudinal tracheids, indicating an open structure of both simple and half-bordered pit membranes. The pattern of penetration was similar in heartwood and sapwood except that a much higher pressure was required for the former. In the heartwood the latewood was more easily penetrated than the earlywood, due to less pit aspiration.

Bailey and Preston (1969) treated small cubical specimens of *Pseudotsuga menziesii* with silver nitrate solution and a precipitating agent. Ultrathin sections were observed by transmission electron microscopy. The distribution of silver deposits indicated flow between longitudinal tracheids through bordered pit pairs. No significant passage was detected between longitudinal tracheids and ray parenchyma or between ray parenchyma cells, indicating relatively impermeable simple and half-bordered pits in disagreement with Wardrop and Davies' results with *Pinus radiata*. Silver deposits were observed in the cell walls of air-dried and green sapwood, suggesting the presence of a capillary system accessible to polar liquids. The cell walls of dried and green heartwood specimens were not similarly penetrated, but behavior similar to that of the sapwood was achieved after extraction or delignification. No dimensions were given for the cell-wall capillaries, but measurements from the electron micrographs indicate a maximum size of silver desposits of approximately 80 nm.

Liese and Bauch (1967a) attributed the poor treatability of *Picea abies* wood to the low permeability of the rays and particularly to the small volumetric fraction of ray tracheids. In *Pseudotsuga menziesii* similar refractory behavior is caused by heavily encrusted pits between the ray tracheids, in addition to a low fraction of this cell type. On the other hand, the superior treatability of *Pinus sylvestris* may be attributed to a high fraction of ray tracheids with an open pit structure. These results were corroborated by Banks (1970), who measured a much higher radial than tangential permeability in the wood of *Pinus sylvestris*. Similar results by several investigators were reported by Comstock (1970) in which ratios of radial to tangential permeability extended from 5 to 37 for pines, 0.04 to 10 for spruces, and 0.03 to 0.7 for *Sequoia sempervirens, Juniperus virginiana,* and *Pseudotsuga menziesii* heartwood. These groups are arranged in the order of decreasing treatability and decreasing longitudinal permeability, thus emphasizing the importance of permeable ray tissue in the preservative treatability of wood.

In summary, it should be pointed out that the microscopic flow studies which have been reviewed have all utilized small wood specimens and there are difficulties in predicting the treatability of large specimens from these results. For example, there is general agreement that the principal longitudinal flow path is from tracheid to tracheid through the bordered pit pairs. Also there is usually an effective flow path from ray tracheids to longitudinal tracheids. In the treatment of long, slender poles, the longitudinal penetration may be negligible relative to the radial. Radial penetration between longitudinal tracheids would be very limited because the pitting is concentrated on the radial surfaces which would promote tangential flow. Therefore it is probable that radial penetration through the rays, and then to the longitudinal tracheids is the most important route through which wood preservatives enter the sapwood of a long specimen. In regard to the penetrability of polar and nonpolar liquids, Nicholas and Siau (1973) have pointed out that the flow rate of nonpolar liquids such as oils is significantly greater than that of polar liquids of the same viscosity. The effect is attributed to a frictional drag due to hydrogen bonding to cell-wall components during flow through small capillaries. However, there is also evidence that polar liquids such as water-borne preservative solutions are capable of penetrating the cell wall, while nonpolar liquids may only penetrate the lumens of the cells. Finally, sapwood is nearly always more permeable than heartwood due to the effect of pit aspiration and incrustation in the latter, and latewood is generally more permeable than earlywood because of decreased pit aspiration during drying.

2.7 Structure of Hardwoods

The structure of a typical hardwood is illustrated in Fig. 2.21 with designations of the principal cell and tissue types including the vessels, fibers, and rays in both earlywood and latewood. Hardwoods are much more varied and complex than softwoods and are subdivided into diffuse-porous and ring-porous types. The former contain vessels which are relatively uniform in size and distribution, as illustrated by the scanning electron micrograph of *Liriodendron tulipifera* in Fig. 2.22. Scalariform perforation plates are seen between the vessel elements as well as extensive intervessel pitting. The vessels are approximately 50 to 100 µm in diameter and are surrounded by relatively thick-walled fiber cells. Multiseriate rays are seen in axial view on the cross-section surface and in transverse section on the tangential sur-

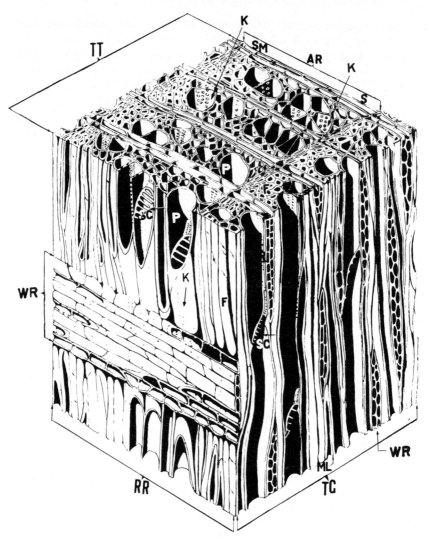

Fig. 2.21. Gross structure of a typical hardwood. Plane *TT* cross section; *RR* radial surface; *TG* tangential surface. The vessels or pores are indicated by *P*, and the elements are separated by scalariform perforation plates, *SC*. The fibers, *F*, have small cavities and thick walls. Pits in the walls of the fibers and vessels, *K*, provide for the flow of liquid between the cells. The wood rays are indicated at *WR*. *AR* indicates one annual ring. The earlywood (springwood) is designated *S*, while the latewood (summerwood) is *SM*. The middle lamella is located at *ML*. (MacLean 1952)

face. A typical ring-porous structure of an oak is shown in Fig. 2.23 with very large vessel elements from 200 to 350 μm in diameter in the earlywood, separated by simple perforation plates. Latewood vessels vary in diameter from 30 to 100 μm. Multiseriate rays are seen on the radial surface, and some rays are also evident on the cross section. The vessels are surrounded by vasicentric tracheids and thick-walled fiber cells. Many hardwoods have porous structures which are intermediate between these diffuse-porous and ring-porous types.

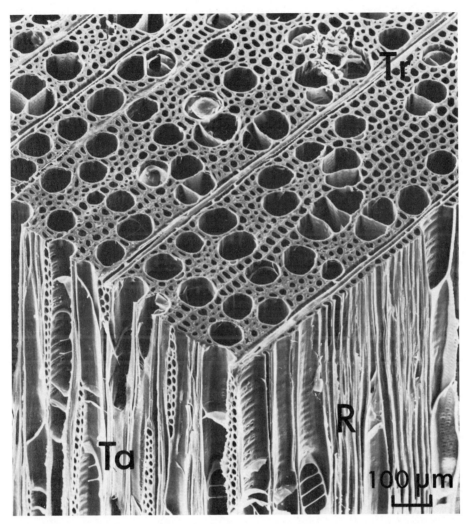

Fig. 2.22. Scanning electron micrograph of the diffuse-porous hardwood *Liriodendron tulipifera*, showing vessels with diameters from 50 μm to 100 μm on the transverse surface *(Tr)* with scalariform perforation plates between the vessel elements visible on both the tangential *(Ta)* and the radial *(R)* surfaces. Thick-walled fibers are located between the vessels with a diameter from 10 μm to 25 μm. Pitting occurs between vessels, vessels and rays, and vessels and fibers on the radial surface of the cells. Multiseriate rays are seen in transverse section on the tangential surface and in longitudinal section on the transverse surface. (Courtesy of the N.C. Brown Center for Ultrastructure Studies)

The prosenchyma tissue of hardwoods includes vessels (called pores in cross section), and fibers. The latter are usually thick-walled and sparsely pitted and include fiber tracheids with bordered pits and libriform fibers with simple pits. Also included in the conductive tissue are vasicentric tracheids, which are short, fibrous cells with bordered pits, usually associated with earlywood vessels in the Fagaceae and Myrtaceae families. Vascular tracheids, which resemble small vessel elements

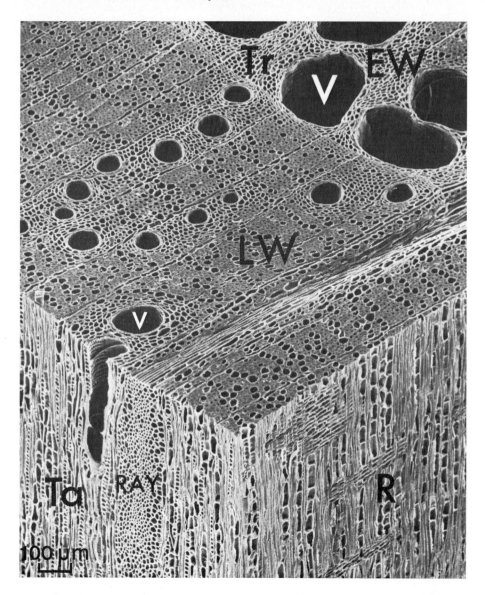

Fig. 2.23. Scanning electron micrograph of the ring-porous hardwood *Quercus rubra*, showing earlywood *(EW)* vessels *(V)* with a diameter between 200 μm and 350 μm and latewood *(LW)* vessels with a diameter from 30 μm to 100 μm. Simple perforation plates are present between the vessel elements. Nearly one complete annual ring can be seen on the transverse surface *(Tr)*. Vasicentric tracheids are associated with both the earlywood and the latewood vessels. A wide ray *(RAY)* and several narrow, uniseriate rays are seen on the tangential surface *(Ta)* Narrow rays are visible in longitudinal view on the radial surface *(R)*. A large fraction of the transverse surface consists of thick-walled fibers with a small diameter. (Courtesy of the N.C. Brown Center for Ultrastructure Studies)

but with imperforate ends and bordered pit pairs, are found in few species and in a small volume fraction. The parenchyma cells include ray parenchyma and the epithelial cells which surround the vertical and horizontal gum canals. The gum canals are not cells but intercellular spaces of low volumetric fraction.

Panshin and de Zeeuw (1980) give the volumetric composition of *Liquidambar styraciflua*, a typical diffuse-porous hardwood as:

Vessels	55%
Fiber tracheids	26%
Longitudinal parenchyma	1%
Wood rays	18%

The vessels of diffuse porous woods usually occupy a larger volumetric fraction than in ring-porous woods with a range of approximately 20% to 60%. The diameters are relatively uniform within a specimen and may vary from 20 µm to 100 µm in most temperate-zone hardwoods. Larger vessels may be found in some tropical hardwoods. An average concentration of vessels, as viewed in cross section, is 15,000/cm^2. In ring-porous woods the vessels occupy relatively low volumetric fractions of approximately 5% to 25% with the majority consisting of large earlywood vessels. The diameters of earlywood vessels are generally between 50 and 400 µm, while the latewood vessels may have diameters between 20 and 50 µm.

Typical hardwood elements are illustrated in Fig. 2.24 with a softwood tracheid for comparison. It is clear from this that vessel elements are relatively short. They are aligned end-to-end with perforation plates between them. The *simple* perforation plate has a single opening almost as large as the vessel lumen (Fig. 2.23), and the *scalariform* perforation plate has several openings separated by thin, transversely oriented bars (Fig. 2.22). The resistance to flow of either type of perforation plate is low because the openings are large and the plates are thin. Therefore the vessels behave as long, open tubes or capillaries. The vessels, composed of many elements, are not of infinite length, furthermore, there is frequent branching, and there is communication between vessels through intervessel pits. In the oaks there is communication between vessels trough vasicentric tracheids (Wheeler and Thomas 1981). The pits leading to fibers or tracheids are usually bordered, while those leading to parenchyma cells are usually half-bordered or simple.

Tyloses may greatly increase the resistance to flow through heartwood and sometimes sapwood vessels of many species. Tyloses are cellular membranes which enter the vessels from adjacent parenchyma cells through pit pairs. They account for the extremely low permeability of the white oaks. Gummy substances such as the reddish deposits in honeylocust (*Gleditsia triacanthos*) and the black material in ebony may also occlude the vessels. Chalky deposits may be found in mahogany and teak.

Fibers can account for 20% to 70% of the wood volume, depending on the species. They include fiber tracheids and libriform fibers, with an example of the latter illustrated in Fig. 2.24. One or the other, or both types, are found in many species. They are generally much smaller than softwood tracheids, are thicker-walled, and have fewer and smaller pits, mostly located on the radial surface. Panshin and de Zeeuw (1980) give an range of lengths from 600 to 2,300 µm, with an average of approximately 1,500 µm and a length-to-diameter ratio of 100. Fluids are able to

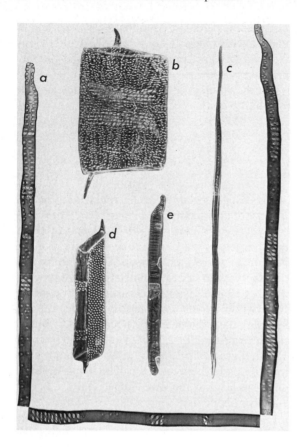

Fig. 2.24. Comparison of the size and shape of softwood and hardwood elements. The softwood tracheid (**a**) has a length of 3.5 mm and a diameter of 40 µm. Most of its pits are at the tapered ends which overlap with adjacent tracheids. Vessel elements (**b, d, e**) vary in diameter and are all short. Most of the pits are on the tangential surface in **d**. The libriform fiber (**c**) is very thin and relatively short. (Photomicrograph. Kollmann and Côté 1968)

reach fiber lumens from vessels and parenchyma cells through pit pairs, although the permeability and treatability of fiber tissue is poor. For example, a small fraction of the fibers of the oaks is generally filled when the wood is subjected to normal pressures in wood-preservative treatments.

The longitudinal parenchyma tissue may make up 1% to 18% of the volume of domestic woods and it is usually not easily penetrated. In tropical woods, however, more than 50% of the volume may be longitudinal parenchyma which is frequently more permeable than the fibers. Ray parenchyma is present to the extent of 5% to 33% in domestic woods with an average of 17%. Although ray parenchyma is frequently impermeable, it may in some instances provide an effective radial path for penetration of fluids as in softwoods.

2.8 Hardwood Pitting

Examples of hardwood pitting are illustrated in Figs. 2.25 to 2.27. There are generally no tori in hardwood pits and the membranes are continuous across the entire pit chamber and consist of primary-wall material of random orientation, as shown

Fig. 2.25. Bordered pit between a vessel and a vasicentric tracheid in *Quercus falcata* with a pit membrane consisting of randomly oriented cellulose microfibrils. (Courtesy of E.A. Wheeler)

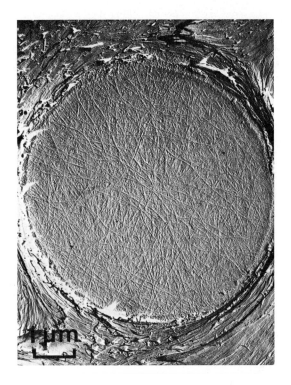

in Fig. 2.25. Hardwood pits are generally smaller than those in softwoods, with the chamber diameters of those in Figs. 2.26 and 2.27 being approximately 6 µm. Although hardwood pit membranes usually have no apparent openings comparable with those between the microfibrillar strands in the margo of softwood membranes, there is ample evidence that openings are present. For example, hardwoods are permeable to gases and liquids in the tangential and radial directions. Bonner and Thomas (1972) have published electron micrographs which reveal openings up to 100 nm in the intervessel pit membranes of *Liriodendron tulipifera*. Smulski (1980) has presented a scanning electron micrograph in which an alkyd resin replica reveals the flow of resin between procumbent ray cells of *Tilia americana* with an estimated diameter of approximately 70 nm. Murmanis and Chudnoff (1979) treated specimens of *Fagus sylvatica* with a silver nitrate solution and a precipitating agent for tracing flow paths and found silver deposits in the vessel pit membranes with diameters of 70 nm or less. On the other hand, India ink particles and carbon particles, both with diameters of 35 nm, did not penetrate the pit membranes. This was attributed to their nonpolar nature. It may be concluded from this that polar solvents enter the cell wall more easily than nonpolar substances and therefore will result in a larger diameter determination for a similar membrane. Cronshaw (1960) found that colloidal gold and carbon particles with diameters of 64 nm did not pass through the pit membranes of *Eucalyptus regnans* wood. Rudman (1965) noted that both polar and nonpolar solvents penetrated the pit membrane of this species. Harada (1962) observed openings with diameters between 80

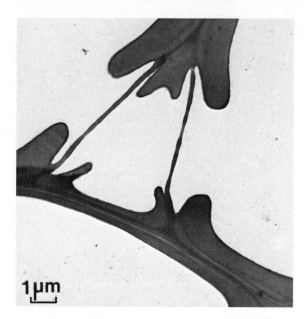

Fig. 2.26. Intervessel pitting in *Quercus rubra*. Note the absence of any tori. Inside diameter of chamber: 6 μm, thickness of membrane, 0.15 μm. Transverse section. TEM by W.A. Côté. (Siau 1971)

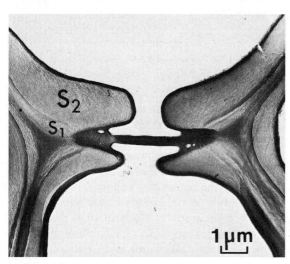

Fig. 2.27. Intervessel pitting in *Quercus rubra*, showing cell wall layers. Inside diameter of chamber: 6 μm, thickness of membrane: 0.3 μm. Transverse section. TEM by W.A. Côté. (Siau 1971)

and 170 nm in the membranes of simple pit pairs between vertical parenchyma cells of *Fagus crenata* wood.

It is clear from these results that there is a large variation in the sizes of hardwood pit membrane pores as there is in softwoods, and that the values obtained in measurements depend on the method used. No minimum values appear in the literature but Murmanis and Chudnoff (1979) reported very small openings up to the maximum of 70 nm. If a minimum of 5 nm is assumed, a range of 5 to 170 nm is included in these references, with a logarithmic mean of 30 nm which is approximately one order of magnitude smaller than the value for softwoods.

The thickness of the pit membranes depicted in Fig. 2.26 is 0.15 µm and that in Fig. 2.27 is approximately 0.3 µm. Harada (1962) gives a range of 0.18 to 0.27 µm for the bordered pit membranes of *Fagus crenata* wood. These values indicate a probable range of 0.1 to 0.3 µm, which is similar to that in softwoods. Harada also measured 0.40 µm for a simple pit membrane between vertical parenchyma cells, indicating thicker membranes in this pit type which is also the case in softwoods.

2.9 Microscopic Studies of Flow in Hardwoods

Different results may be expected from polar and nonpolar liquid penetrations. Therefore the investigation of Behr et al. (1969), using creosote and pentachlorophenol in gas oil, will be discussed first. Four domestic hardwoods were impregnated with oily preservatives using an empty-cell process. Little difference was observed between the distribution of creosote and pentachlorophenol. The vessels were found to be the principal flow paths with more latewood than earlywood vessels filled in both the ring-porous and diffuse-porous woods. This was probably due to the fact that smaller vessels were able to hold the oil due to greater capillary forces. The fibers of most of the specimens were filled, indicating flow from the vessels through pits. No significant difference was found between the penetration of earlywood and latewood fibers such as was found in the coniferous woods. Vasicentric tracheids of *Quercus rubra* were easily penetrated. Vertical parenchyma cells were usually not filled, in disagreement with the findings of Wardrop and Davies (1961) for tropical woods. Very little penetration was observed in the multiseriate rays of *Fagus americana*, *Quercus rubra*, and *Carya* sp., although the uniseriate rays usually contained oil. The rays of the hardwoods generally made less of a contribution to the overall flow than those of the softwoods similarly investigated, despite the generally higher volumetric fraction of rays in hardwoods. Côté (1963) stated that the effect of rays on fluid conduction is much more variable in hardwoods than in softwoods. The relative role of rays in softwoods and hardwoods may also be assessed on a basis of their contribution to permeability. High ratios of radial to tangential permeability were found in some softwoods, particularly those of high treatability (Sect. 2.6). Comstock (1975 unpublished results) has reported low ratios between 0.3 and 2.0 for eight hardwood species, indicating little difference between radial and tangential permeability in general. Therefore the contribution of the rays to the radial permeability of hardwoods is about equal to that in the tangential direction resulting from the pits on the radial surfaces of the fibers.

Rudman (1965) treated specimens up to 2.5 m in length of four species of eucalypt wood with five nonpolar solvents including creosote and four polar solvents. He found very little penetration of the ray parenchyma except by hot polar solvents. Both the polar and nonpolar liquids entered the wood primarily through the vessels and then penetrated the vasicentric tracheids, vertical parenchyma, and finally the fibers through the pit pairs. Many of the vessels in the long specimens were penetrated via thin radial checks. Both polar and nonpolar liquids passed through the pit membranes.

Wardrop and Davies (1961) treated small specimens of *Eucalyptus regnans* sapwood with aqueous salt solutions and suitable precipitating agents for the observation of flow paths. Penetration was primarily through the open vessels, then through rays, vertical parenchyma, and fibers through pit pairs. Heartwood gave similar results, except that some of the vessels were closed by tyloses. Other hardwoods with well-defined bands of vertical parenchyma exhibited similar behavior along with complete penetration of the thin-walled parenchyma bands.

Rudman (1966) impregnated small specimens of *Eucalyptus obliqua* and *Eucalyptus maculata* alternatively with aqueous silver nitrate and a precipitating agent and a copper-chromium-arsenic preservative solution. The presence of deposits within the cell walls was interpreted as evidence of cell-wall capillaries. The size, as determined by measurements from electron micrographs, was in the range of 10 to 70 nm, with the possibility of some capillaries smaller than 10 nm. In a similar investigation, some pit membranes and cell walls were not penetrated, probably due to extractives and cytoplasmic debris. Murmanis and Chudnoff (1979) treated small wafers of *Fagus sylvatica* and *Betula papyrifera* wood with aqueous solutions of silver nitrate and copper sulfate and suitable precipitating agents. Flow was restricted to the transverse direction. Bulk flow was observed from the rays to the vessels via pit pairs. Although the copper solution did not penetrate the cell walls, the presence of silver deposits indicated cell-wall capillaries with sizes up to 70 nm in the vessels and rays and up to 130 nm in the fibers. Both this study and that of Rudman (1966) indicate dual paths of penetration by polar liquids; namely through the pit-membrane openings and through the cell wall. In woods of high permeability, the pit path will predominate, while in woods of very

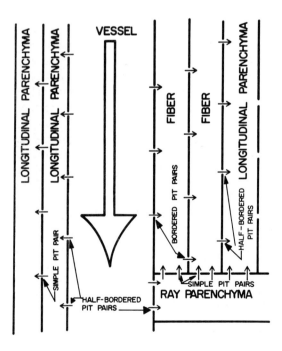

Fig. 2.28. Generalized flow model for hardwoods. (Siau 1971)

low permeability the cell-wall path may be the most significant. In this regard, cell-wall permeability has been measured by Palin and Petty (1981) (Chap. 3.17).

In summary, the vessels are generally the most permeable flow path in hardwoods when they are not occluded by tyloses. Liquids then flow from vessels to fibers, vasicentric tracheids when present, vertical parenchyma, and rays. Rays are generally much less easily penetrated than in softwoods; therefore, the longitudinal flow through the vessels is important, even in long specimens due to the branching of vessels and the role of intervessel pits. Radial drying checks may also be important pathways to vessels in long specimens. An alternative path of polar liquids is through the cell wall when it is not incrusted. A generalized flow model for hardwoods is illustrated in Fig. 2.28. As in softwoods, sapwood is generally more permeable than heartwood. There is no evidence of decreased earlywood permeability as there is in softwoods because the lack of tori prevents pit aspiration in the former. In fact the earlywood of ring-porous woods is usually much more permeable than that of diffuse-porous types. Thus the earlywood of hardwoods is frequently more treatable than the latewood.

2.10 Chemical Composition of Normal Wood

2.10.1 Cellulose

Cellulose, which is the most abundant organic chemical on earth, occurs in all land plants and always in a fibrillar form. The amount of cellulose present in normal wood is remarkably constant, namely 42% ± 2% in both softwoods and hardwoods. The chemical composition of a normal wood from a softwood and hardwood species, excluding the extractives, is shown in Table 2.2. Cellulose is a linear, (1→4)-linked glucan, consisting of β-D-glucopyranose residues in a chair conformation. Every glucose residue is turned over 180° with respect to its neighbors, and each has one primary and two secondary hydroxyl groups, as shown in Fig. 2.29.

Table 2.2. Chemical composition of normal wood of *Pinus strobus* and *Betula papyrifera*. All values in percent of oven-dry, extractive-free wood

Component	*Pinus strobus*	*Betula papyrifera*
Cellulose	42	42
Glucomannan		3
Galactoglucomannan	18	
Glucuronoxylan		30
Arabinoglucuronoxylan	11	
Lignin	28	24
Other hemicellulose, pectin, and ash	1	1

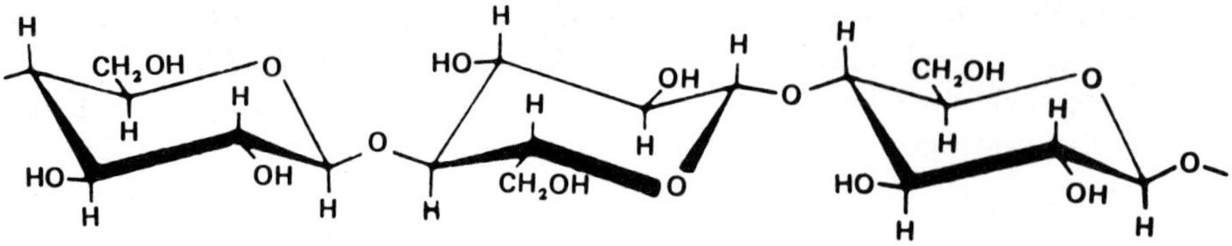

Fig. 2.29. Chemical formula of cellulose. (Blackwell 1982)

Water can be bound by hydrogen bonds to the hydroxyl groups, forcing adjacent chains apart, thus causing swelling of the fiber. Increased swelling results from the action of aqueous alkali.

Wood cellulose in the secondary wall has an average degree of polymerization (DP) of 10,000, corresponding to a length of about 5 µm (Goring and Timell 1962) and is probably monodisperse (Marx-Figini and Schultz 1966). Primary-wall wood cellulose, on the other hand, has a DP of only 2,000 to 4,000 and is polydisperse (Simson and Timell 1978b).

Wherever it occurs in nature, cellulose is partly crystalline and in wood to the extent of 50% to 60%. The unit cell of crystalline native cellulose (Cellulose I) is shown in Fig. 2.30. The cell is monoclinic with all chains oriented in the same direction (parallel) (Gardner and Blackwell 1974; Sarko and Muggli 1974). The length of the repeating unit along the c-axis (cellobiose) is 10.3 Å (1.03 nm). There are strong hydrogen bonds both between adjacent glucose residues within each chain and between neighboring chains in the a-c plane. There are no hydrogen bonds between the chains in the b-c plane. Cellulose is organized in both a chain and a layer lattice, which is impermeable to water. When water enters between the chains of the amorphous or paracrystalline regions, the first water molecules become strongly bound to the hydroxyl groups of cellulose, pushing the chains apart and causing swelling. The length of the chains is not affected.

The crystalline structure of Cellulose I is irreversibly destroyed by swelling with strong alkali or regeneration from solution. The resulting Cellulose II, which is thermodynamically more stable than Cellulose I, has a different unit cell with antiparallel chains, which are more strongly hydrogen-bonded to one another. The reason why native cellulose has the crystalline structure of the less stable Cellulose I is associated with the manner in which cellulose is synthesized in nature by mobile enzyme complexes on and within the plasma membrane of the cell (Brown 1982).

Wood cellulose is organized in microfibrils with a width of 1 to 2 nm in the primary and about 10 nm in the secondary wall. Ultrasonic treatment causes formation of so-called elementary fibrils from secondary wall microfibrils, but it is uncertain if these constitute the ultimate biological unit of cellulose as has been suggested. The fact that the chains in Cellulose I are parallel excludes any possibility of chain folding in native cellulose. The arrangement and possible size of the crystallites in native cellulose are schematically illustrated in Fig. 2.31. Cellulose is largely responsible for the tensile strength of wood and also contributes to the water adsorption of wood through its numerous hydroxyl groups.

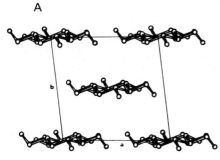
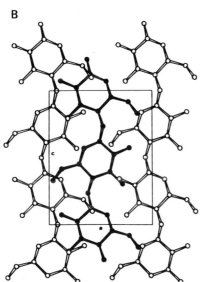
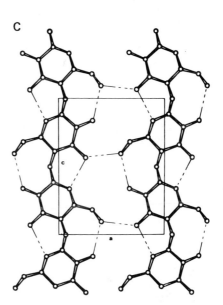

Fig. 2.30. Crystal structure of native cellulose (Cellulose I). **A** ab-projection; **B** ac-projection; **C** network of hydrogen bonds in the ac-plane. (Gardner and Blackwell 1974)

2.10.2 Hemicelluloses

2.10.2.1 Introduction

Hemicelluloses are polysaccharides associated with cellulose and lignin in the cell wall of land plants (Timell 1964, 1965). Their much lower molecular weight distinguishes them from the cellulose and pectin. Primary walls consist of cellulose (25%), pectin (40%), hemicelluloses (25%), and protein (10%). A major hemicellulose is a galactoxyloglucan not present in the secondary wall (Bauer et al. 1973; Simson and Timell 1978a).

Pectin is a complex, highly branched polysaccharide. Its major constituent is (1→4)-linked α-D-galacturonic acid, partly present as a methyl ester and partly as a calcium salt. Native pectin also contains other sugar residues, namely L-arabinose, D-galactose, L-rhamnose, and L-fucose, all except for the rhamnose attached as side chains to the galacturonan backbone. The molecular weight of pectin is of the order 10^6. Pectin has only on a few occasions been observed to be present in the form of microfibrils.

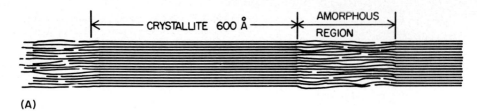

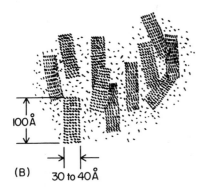

Fig. 2.31 A, B. Schematic structure of a portion of a cellulose microfibril with crystallites and amorphous (disordered) regions according to the fringe micellar theory (A) and a transverse view of several crystallites and amorphous regions (B). (Siau 1971)

2.10.2.2 Softwood Hemicelluloses

The predominant hemicellulose in softwood is a family of galactoglucomannans, which together account for 15% to 20% of this type of wood. They consist of a main chain of (1→4)-linked β-D-glucopyranose and β-D-mannopyranose residues, some of which have a single side chain of α-D-galactopyranose attached to their 6-position, as indicated in Fig. 2.32. The glucose to mannose ratio is about 1:3 but the ratio of galactose to glucose can vary from 1:1 to 1:10. Acetyl groups are also attached to some of the residues in the glucomannan backbone. The galactoglucomannans are present in an amorphous state in wood.

Softwoods also contain about 10% of a xylan with a backbone of (1→4)-linked β-D-xylopyranose residues with two types of side chains, namely a 4-O-methyl-α-D-glucuronic acid residue attached to C-2 and an α-L-arabinofuranose unit linked to C-3. There is one acid side chain per 5 xylose and one arabinose per 7 xylose residues, as indicated in Fig. 2.33. There are no acetyl groups.

Arabinogalactan is a polysaccharide unique to members of the genus *Larix*, where it occurs in amounts ranging from 5% to 30%. Unlike all other wood hemicelluloses, larch arabinogalactan is a highly branched polymer. It consists of a main chain of (1→3)-linked β-D-galactopyranose residues, all of which carry side

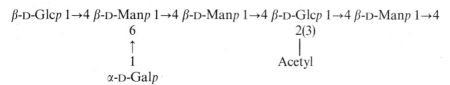

Fig. 2.32. Schematic structure of a softwood galactoglucomannan

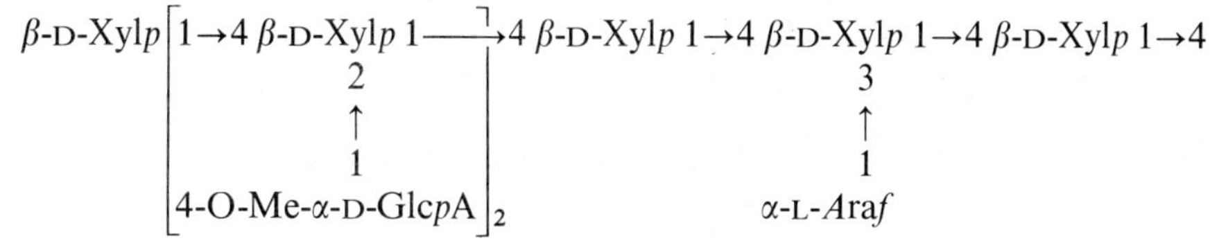

β-D-Xyl*p* $\begin{bmatrix} 1\to 4\ \beta\text{-D-Xyl}p\ 1 \\ 2 \\ \uparrow \\ 1 \\ 4\text{-}O\text{-Me-}\alpha\text{-D-Glc}p\text{A} \end{bmatrix}_2$ →4 β-D-Xyl*p* 1→4 β-D-Xyl*p* 1→4 β-D-Xyl*p* 1→4
                                                                                                    3
                                                                                                    ↑
                                                                                                    1
                                                                                                    α-L-Ara*f*

Fig. 2.33. Schematic structure of a softwood arabinoglucuronoxylan

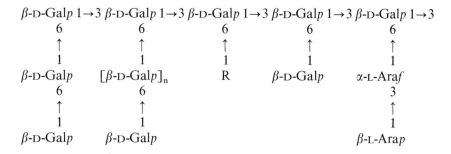

R = α-L-Ara*f* or β-D-GlcpA

Fig. 2.34. Schematic structure of a larch arabinogalactan

chains of variable length, composed largely of (1→6)-linked β-D-galactopyranose, β-D-glucuronic acid, and L-arabinose residues in two different forms, as indicated in Fig. 2.34. Contrary to the other hemicelluloses, larch arabinogalactan is extracellular and occurs only in the lumens of the tracheids and ray cells within the heartwood region. It is also unique in being composed of two polymers of similar structure but different size. The major portion has a molecular weight of 70,000 and the lesser 12,000 (Simson et al. 1968). Both are subjected to a slow, acid hydrolysis in the living tree.

2.10.2.3 Hardwood Hemicelluloses

The major hemicellulose in all hardwoods is an acidic xylan, present in an amount of 25% ± 5% by weight of the extractive-free wood. The xylan content can reach 35% in a few species, for example in some birches. The xylan in hardwoods consists of a linear, main chain of (1→4)-linked β-D-xylopyranose residues. Single-unit side chains, consisting of a (1→2)-linked 4-*O*-methyl-α-D-glucuronic acid residue, are randomly distributed along the xylan backbone, usually with an average number of one side chain per 10 xylose units. There are also 7 acetyl groups per 10 xylose residues attached to the hydroxyl groups of the xylan chain. A schematic formula of this xylan is shown in Fig. 2.35. In addition, it has recently been found that an L-rhamnose and a D-galacturonic acid residue occur near the reducing end of a birch xylan (Johansson and Samuelson 1977). Hardwood xylans have a DP of only 200. They are amorphous in wood but can be induced to crystallize after some of the side chains have been removed. There is good evidence that the xylan macro-

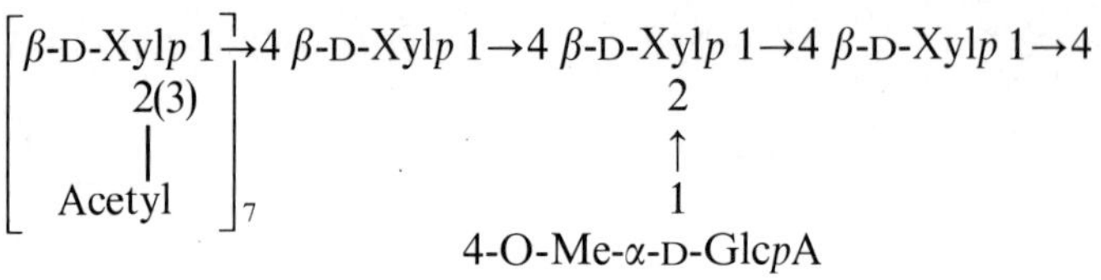

$$\begin{bmatrix} \beta\text{-D-Xyl}p\ 1\xrightarrow{}4\ \beta\text{-D-Xyl}p\ 1\rightarrow 4 \\ 2(3) \\ | \\ \text{Acetyl} \end{bmatrix}_7 \quad \begin{array}{c} \beta\text{-D-Xyl}p\ 1\rightarrow 4\ \beta\text{-D-Xyl}p\ 1\rightarrow 4 \\ 2 \\ \uparrow \\ 1 \\ \text{4-O-Me-}\alpha\text{-D-Glc}p\text{A} \end{array}$$

Fig. 2.35. Schematic structure of a hardwood glucuronoxylan

molecules are oriented parallel with the cellulose microfibrils (Liang et al. 1960; Page et al. 1976). Like almost all hemicelluloses in land plants, hardwood xylan is never present in the form of microfibrils.

Hardwoods also contain about 5% of a glucomannan of unknown molecular size. This hemicellulose consists of (1→4)-linked β-D-glucopyranose and β-D-mannopyranose residues, most frequently in a 1:2 ratio.

2.10.3 Lignins

Lignins are three-dimensional polymers composed of phenylpropane units with a somewhat different composition in softwoods and hardwoods (Freudenberg and Neish 1968; Sarkanen and Ludwig 1971; Adler 1977). They encrust the intercellular space and any openings in the cell wall after the cellulose and hemicelluloses have been deposited. Most normal softwoods contain 30%±4% lignin and hardwoods of the temperate zone 25%±3%. Lignins are polymerization products of p-coumaryl, coniferyl, and sinapyl alcohols (Fig. 2.36).

Almost all conifer species have a quaiacyl lignin composed largely of phenyl propane units with one phenolic oxygen and one methoxy group, derived from coniferyl alcohol. A typical softwood lignin contains the following functional groups per 100 building units: methoxyl (90 to 95), phenolic hydroxyl (20), phenolic ether (80), aliphatic hydroxyl (90), benzyl alcohol or ether (40), and car-

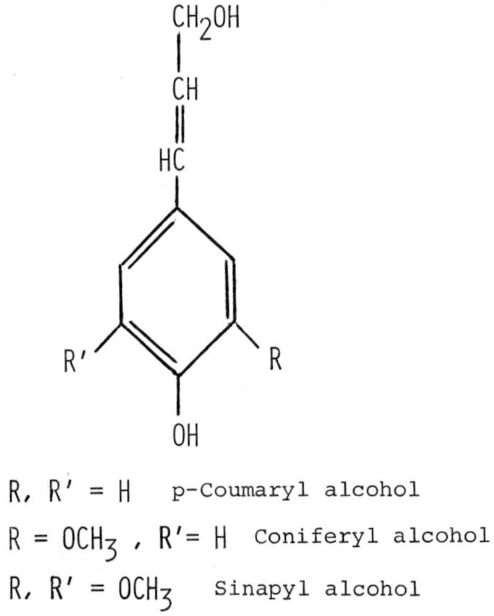

R, R′ = H   p-Coumaryl alcohol
R = OCH$_3$, R′= H   Coniferyl alcohol
R, R′ = OCH$_3$ Sinapyl alcohol

Fig. 2.36. Structure of p-coumaryl, coniferyl, and sinapyl alcohols

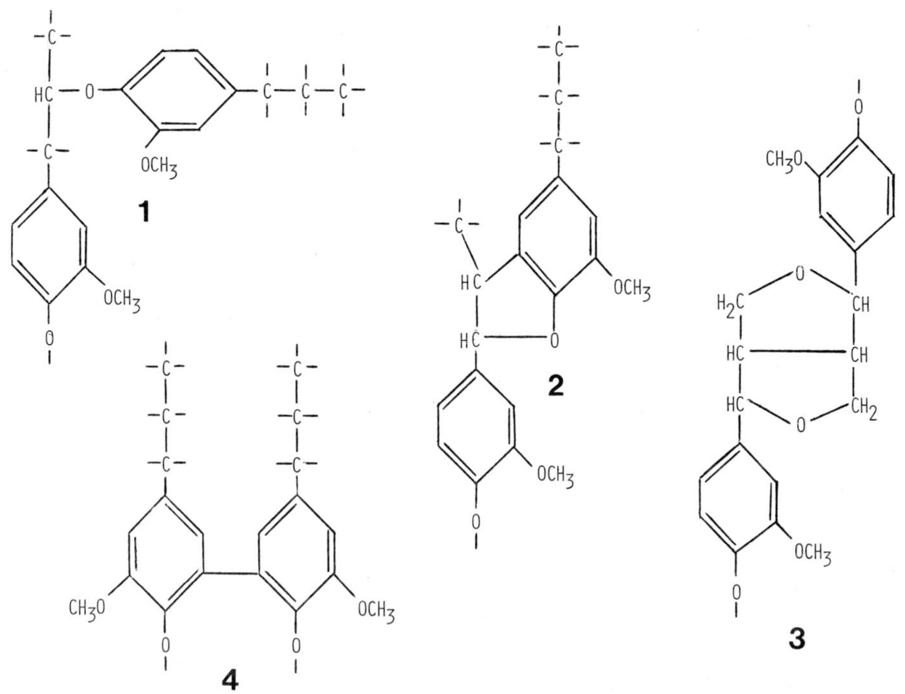

Fig. 2.37. Four major interunit bonds in lignin. *1* arylglycerol-β-aryl ether; *2* phenylcoumaran; *3* pinoresinol; *4* diphenyl

bonyl (20). Approximately one-third of the interunit linkages in lignin are carbon–carbon bonds and the remainder carbon–oxygen. The most important interunit linkage is the arylglycerol β-aryl ether bond (50) (Fig. 2.37). Other major bonds are the phenylcoumaran, the pinoresinol, and the diphenyl types of linkages (each 10). In addition, there are a considerable number of minor linkage types and even a few units that lack a propane side chain. A schematic, two-dimensional formula of a softwood lignin according to Freudenberg (1968) is shown in Fig. 2.38.

Hardwood lignin consists of both guaiacyl and syrinyl units, and has so far generally been considered to be a copolymer of coniferyl and sinapyl alcohols in ratios that vary widely between different species. The proportion of interunit linkages of the β-aryl ether type is higher in hardwood than in softwood lignins. Because of the presence of an additional methoxyl group in the syringyl units, a lignin in hardwood is less highly condensed than that in softwood.

All lignins absorb in the ultraviolet region but a softwood lignin far more than a hardwood lignin. The UV absorption has been utilized for estimating the amount and distribution of lignin in wood. Evidence has been adduced in later years that lignin is chemically linked to the polysaccharides in wood, for example to the side chains in the xylans and galactoglucomannans (Eriksson et al. 1980). The axial compressive strength of wood is partly due to its lignin content and partly to the microfibril angle in the S_2 layer. Lignin offers protection to the wood against microbial degradation. It also serves to reduce its hygroscopicity, although lignin, be-

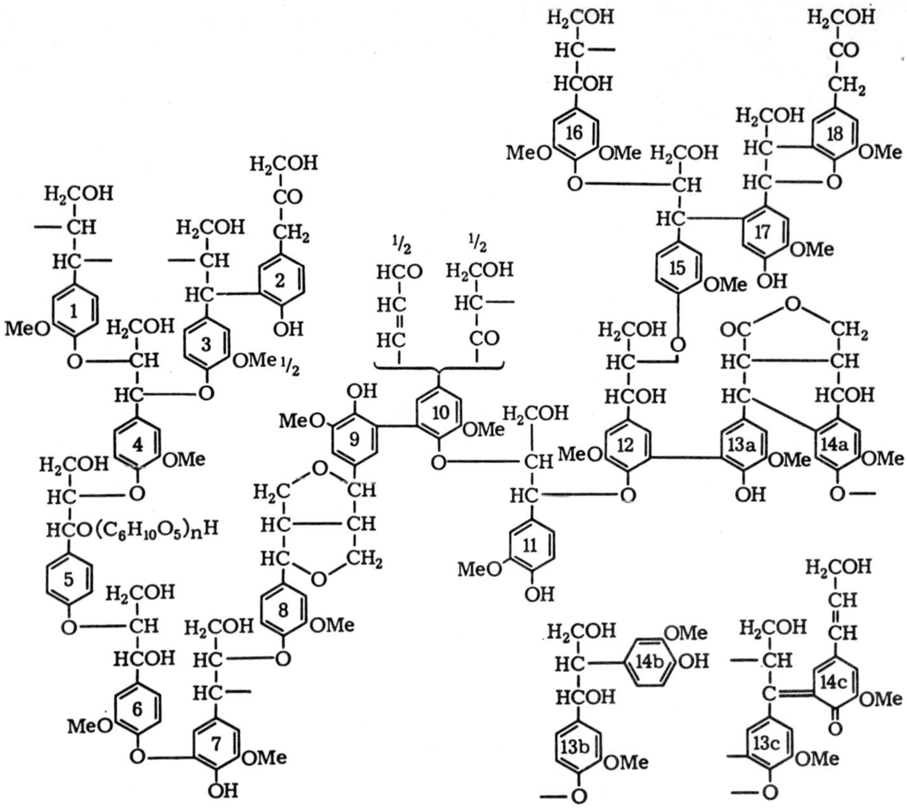

Fig. 2.38. Schematic, two-dimensional formula of a spruce lignin. (Freudenberg 1968)

cause of its numerous hydroxyl groups, is not entirely incompatible with water. Cellulose, and especially the hemicelluloses are, however, far more hygroscopic than lignin.

2.11 Chemical Composition of Reaction Wood

2.11.1 Introduction

Reaction wood, compression wood in conifers and tension wood in angiosperms, differs radically from normal wood, chemically and physically as well as anatomically. The chemical composition of reaction wood of a softwood and a hardwood is shown in Table 2.3.

2.11.2 Compression Wood

Compression wood has only 30% cellulose compared to 42% in a normal softwood. Its lignin content can be as high as 40% (Timell 1981, 1982). The cellulose in compression wood is less crystalline than in normal wood (Tanaka et al. 1981). Compression wood contains two hemicelluloses present in only trace amounts, if

Table 2.3. Chemical composition of reaction woods of *Pinus strobus* and *Betula papyrifera*. All values in percent of oven-dry, extractive-free wood

Component	*Pinus strobus*	*Betula papyriféra*
Cellulose	30	50
Laricinan	3	
Glucomannan		2
Galactoglucomannan	9	
Galactan	10	8
Xylan	9	22
Lignin	38	17
Other hemicelluloses, pectin, and ash	1	1

$$\beta\text{-}{\rm D}\text{-}{\rm Gal}p\ 1{\to}4\ \beta\text{-}{\rm D}\text{-}{\rm Gal}p\ 1{\to}4\ \beta\text{-}{\rm D}\text{-}{\rm Gal}p\ 1{\to}4\ \beta\text{-}{\rm D}\text{-}{\rm Gal}p\ 1{\to}4$$
$$6$$
$$\uparrow$$
$$1$$
$$\beta\text{-}{\rm D}\text{-}{\rm Gal}p{\rm A}$$

Fig. 2.39. Schematic structure of a compression wood galactan

at all, in normal softwood. One is a (1→3)-linked β-D-glucan, referred to as laricinan (Hoffmann and Timell 1970). It amounts to only 2% to 4% of the wood. Approximately 10% of compression wood is a (1→4)-linked β-D-galactan with a few β-D-galacturonic acid side chains attached to position 6 of the main chain, as shown in Fig. 2.39 (Bouveng and Meier 1959; Schreuder et al. 1966; Jiang and Timell 1972). Compression wood contains the same amount of xylan as normal wood but only half as much galactoglucomannan.

The lignin in compression wood is a copolymer of p-coumaryl and coniferyl alcohols (Erickson et al. 1973; Sakakibara 1977). It is more highly condensed than a normal softwood lignin because of its lower methoxyl content and has fewer β-aryl ether interunit bonds. It also has a higher proportion of carbon–carbon linkages.

2.11.3 Tension Wood

Because of the presence in tension wood fibers of an unlignified innermost layer, the so-called G-layer (Fig. 2.7), consisting only of cellulose (Norberg and Meier 1966), tension wood has a much higher cellulose content than normal hardwood. Values over 60% have been reported. The lignin content on a weight basis is correspondingly lower, while remaining the same when based on an individual fiber (Timell 1969).

Tension wood from some (*Betula, Fagus*) but by no means all hardwood species contains a galactan which is still only incompletely known (Meier 1962b; Kuo and Timell 1969). Its main chain consists of (1→4)-β-D-galactopyranose residues with numerous side chains of (1→6)-linked β-D-galactopyranose units. Residues of L-arabinofuranose, L-rhamnopyranose, β-D-glucuronic acid, 4-O-methyl-β-D-glucuronic acid, and α-D-galacturonic acid residues are also present. It is undoubtedly the most complex of all wood hemicelluloses. The guaiacyl-syringyl lignin in tension wood does not seem to differ in any respect from that in a normal hardwood.

2.12 Topochemistry of Wood

Juvenile softwood contains less cellulose and galactoglucomannan and more galactan, xylan, and lignin than mature wood. Similar differences exist between earlywood and latewood. The ray cells in softwood and hardwood contain 35% to 40% lignin. Vessels also have a higher lignin content than fibers. Softwood ray cells, unlike the tracheids and fibers, contain more xylan than galactoglucomannan (Perilä 1961).

The distribution of the polysaccharides within the secondary wall of tracheids and fibers is not known in detail because of a lack of suitable techniques. It is believed that the cellulose and galactoglucomannans reach their highest concentration in the S_2 layer and that in the softwoods the S_3 layer is rich in xylan (Meier 1962a). It has also been found that the outer part of S_1 and S_2 are rich in hemicelluloses (Hoffmann and Parameswaran 1976). The galactan in compression wood tracheids is located in S_1 and the outer portion of the S_2 layer (Timell 1982).

Pit membranes originally consist of two primary walls and the middle lamella between them. During differentiation of tracheids, fibers, and vessels the hemicelluloses and pectin are removed by enzymatic hydrolysis, leaving a cellulose network (Butterfield and Meylan 1982). Where present, the torus consists of either pectin or pectin and cellulose. When sapwood is transformed into heartwood, extraneous substances, and especially polyphenols, are deposited onto the pit membrane but not lignin (Thomas 1975).

The distribution of lignin in wood is now relatively well known thanks to the existence of several excellent methods for lignin localization, such as examination in the transmission electron microscope of wood sections treated with permanganate (Kutscha and Schwarzmann 1975), ultraviolet microscopy (Goring 1981), and energy-dispersive x-ray analysis (Saka and Thomas 1982). According to Goring and his co-workers, 20% to 25% of the total lignin in normal softwoods and hardwoods occurs in the intercellular region and primary wall, while 75% to 80% is present within the secondary wall. The concentration of lignin in softwoods is 50% to 60% in the middle lamella-primary wall and 85% to 100% in the true middle lamella at the cell corners. It is only 20% to 25% in the secondary wall. The situation is similar in hardwoods. Goring and his co-workers have found that the fibers and rays in hardwoods have a wall lignin consisting largely of syringyl units while their middle lamella contains a guaiacyl-syringyl lignin. Vessels, by contrast, have a guaiacyl lignin both in their wall and middle lamella.

In compression wood tracheids less than 10% of the total lignin is located outside the secondary wall, and the average lignin concentration in the intercellular region is 60%, which is less than in normal wood. S_1 and the inner, fissured region of S_2 contain approximately 40% lignin. The outer part of S_2 [$S_2(L)$] has a lignin content of 50% to 60% (Wood and Goring 1971; Timell 1982).

Chapter 3

Permeability

3.1 Introduction

The transport of fluids through wood may be subdivided into two main classifications which will be described in this text. The first is the *bulk flow* of fluids through the interconnected voids of the wood structure under the influence of a static or capillary pressure gradient. This is sometimes designated as momentum transfer because it can be attributed to a momentum-concentration gradient. The second is *diffusion* consisting of two types: intergas diffusion, which includes the transfer of water vapor through the air in the lumens of the cells, and bound-water diffusion, which takes place within the cell walls of wood. Some practical applications of bulk flow are the pressure treatment of wood with liquid preservatives for protection against biological decay or fire, the impregnation of wood with monomers for subsequent in situ polymerization in the manufacture of wood-polymer composites, and the impregnation of wood chips with pulping chemicals. Diffusion, on the other hand, occurs during the air-drying or kiln-drying of wood, in the migration of moisture through the wood members of the exterior walls of a building, and in interior woodwork and furniture in response to seasonal changes in relative humidity. The magnitude of the bulk flow of a fluid through wood is determined by its *permeability*. Diffusion will be discussed in detail in Chapter 6. Another type of diffusion which occurs along with the bulk flow of gases, known as Knudsen diffusion, will be discussed in this chapter.

An understanding of bulk flow may be clarified by a brief discussion of the relationship between porosity and permeability. It will be recalled that porosity is the volume fraction of void space in a solid. Permeability, on the other hand, is a measure of the ease with which fluids are transported through a porous solid under the influence of a pressure gradient. It is clear that a solid must be porous to be permeable, but it does not necessarily follow that all porous bodies are permeable. Permeability can only exist if the void spaces are interconnected by openings. For example, a softwood is permeable because the tracheid lumens are connected by pit pairs with openings in the membranes. If these membranes are occluded or encrusted, or if the pits are aspirated, the wood assumes a closed-cell structure and may have a permeability approaching zero.

3.2. Darcy's Law

The steady-state flow of fluids through wood and other porous solids is described by Darcy's law, which may be stated generally as

$$\text{Conductivity} = \text{Flux} \div \text{Gradient}. \tag{3.1}$$

The conductivity is usually assumed constant in the steady-state laws based upon Eq. (3.1). In Darcy's law the conductivity is called permeability. There are many exceptions to Darcy's law when applied to wood, causing the permeability to become variable. These will be discussed in detail later in this chapter. The assumptions and limitations of Darcy's law are discussed at length by Muskat (1946) and Scheidegger (1974). The principal assumptions are:

1. The flow is viscous and linear. Therefore linear velocity and volumetric flow rate are directly proportional to the applied pressure differential.
2. The fluid is homogeneous and incompressible.
3. The porous medium is homogeneous.
4. There is no interaction between the fluid and the substrate.
5. Permeability is independent of the length of the specimen in the flow direction.

Although these assumptions are violated in many instances when Darcy's law is applied to the flow of gases and aqueous liquids through wood, the basic equation remains a useful relationship between the flow rate and the pressure gradient.

In regard to the first assumption, viscous flow generally occurs when capillary openings are small as they are in wood. In such cases there is a relatively high viscous drag because of the high ratio of surface area to volume. Under these conditions the high flow velocities necessary for turbulence are improbable but nonlinear flow (Siau and Petty 1979) can occur at relatively low velocities where a fluid moves from a large to a small capillary such as from a tracheid lumen to a pit opening. As for the second assumption, liquids are essentially incompressible, but the high compressibility of gases must be accounted for as described below. In regard to (3) and (4), wood has an extremely complex and nonhomogeneous structure. This is especially true of hardwoods. Furthermore, when water flows through wood there are hydrogen-bonding forces exerted by the hydroxyl sorption sites on the cell-wall surface. It is on this basis that Nicholas and Siau (1973) explain the lower permeability of wood to water and aqueous solutions than to nonpolar liquids of the same viscosity.

Darcy's law for liquids may be written as

$$k = \frac{\text{Flux}}{\text{Gradient}} = \frac{Q/A}{\Delta P/L} = \frac{QL}{A\Delta P}, \tag{3.2}$$

where k = permeability, cm^3 (liquid)/(cm atm s), Q = volumetric flow rate, cm^3/s, L = length of specimen in the flow direction, cm, A = cross-sectional area of specimen perpendicular to flow direction, cm^2, ΔP = pressure differential, atm.

It is clear from Eq. (3.2) that permeability is numerically equal to the flow rate through a unit cube of a porous solid with unit pressure differential between two opposite faces. This equation is applicable to a specimen with parallel sides and ends such as that illustrated in Fig. 3.1.

When Darcy's law is applied to gaseous flow, the expansion of the gas as it moves through the specimen causes continuous changes in the gradient and the volumetric flow rate. To account for these, the Darcy equation is written in differential form:

$$k_g = -\frac{Q}{A dP/dx} \quad \text{or} \quad k_g dP = -\frac{Q dx}{A}, \tag{3.3}$$

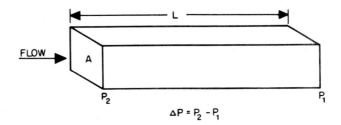

Fig. 3.1. Permeable body illustrating the terms in Darcy's law

where k_g = superficial gas permeability, cm³ (gas)/(cm atm s), x = distance in direction of flow. The term "superficial gas permeability" was originated by Klinkenberg (Scheidegger 1974) because its value generally exceeds liquid permeability when both viscous and slip flows are accounted for. The higher gas permeability can be attributed to Knudsen diffusion, which is discussed later.

The isothermal expansion of an ideal gas due to decreased pressure may be calculated from the general gas law:

$$PV = nRT,$$

where n = moles of gas = w/M_w, w = mass, g, M_w = molecular weight, g/mol, R = universal gas constant = 8.31×10^7 erg/(mol K), T = absolute temperature, K, P = pressure, dyne/cm², V = volume, cm³.

Then

$$V = nRT/P;$$

substituting into Eq. (3.3),

$$k_g P dP = -\frac{nRT}{tA} dL;$$

integrating,

$$k_g \int_{P_1}^{P_2} P dP = -\frac{nRT}{tA} \int_L^0 dx;$$

evaluating the integral,

$$k_g \frac{(P_2^2 - P_1^2)}{2} = \frac{nRTL}{tA};$$

solving for k_g and factoring,

$$k_g = \frac{2nRT}{tA(P_2 - P_1)(P_2 + P_1)}.$$

Since $(P_2 + P_1)/2 = \bar{P}$, the arithmetic average pressure in the specimen and $P_2 - P_1 = \Delta P$, the pressure differential, Darcy's law for gases may be written as

$$k_g = \frac{VLP}{tA\Delta P \bar{P}} = \frac{QLP}{A\Delta P \bar{P}}, \qquad (3.4)$$

(Darcy's law for gases)

where P = pressure at which the flow rate Q is measured.

76 Permeability

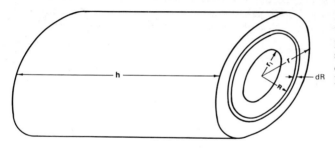

Fig. 3.2. Illustrating the flow through an annulus. Flow is perpendicular to cylindrical surface. P_2 = pressure on outside surface at r, P_1 = pressure at r_i

Darcy's law for an *annulus* may be derived from the equation for a parallel-sided body. Equation (3.2) may be written in differential form as

$$Q = -kA \frac{dP}{dx}.$$

This may be transformed to apply to an annulus by representing A by $2\pi Rh$ where R = variable radius, h = axial length of annulus as represented in Fig. 3.2. Then,

$$Q = \frac{k 2\pi R h dP}{dR}.$$

Separating the variables and integrating,

$$Q \int_{r_i}^{r} \frac{dR}{R} = k 2\pi h \int_{P_1}^{P_2} dP,$$

where P_2 = upstream pressure, P_1 = downstream pressure, r = outside radius, r_i = inside radius.
Then,

$$Q = \frac{2\pi k h \Delta P}{\ln r/r_i}. \tag{3.5}$$

(Darcy's law for liquid flow in an annulus)

By following a derivation similar to that for Darcy's law for gases for a parallel-sided body, the following relationship is obtained.

$$Q = \frac{2\pi k_g h \bar{P} \Delta P}{P \ln r/r_i}. \tag{3.6}$$

(Darcy's law for gas flow in an annulus)

3.3 Kinds of Flow

A brief introduction is presented to the various kinds of flow which can occur in wood. They are called: (a) *viscous* or laminar, (b) *turbulent*, (c) *nonlinear*, due to kinetic energy losses at the entrance of a small, short capillary, and (d) molecular slip flow, sometimes called *Knudsen diffusion*.

Viscosity is internal fluid friction which requires the application of a force to cause one layer of a fluid to flow smoothly past an adjacent layer or to cause one surface to move relative to another when there is a fluid between them. Both gases and liquids possess viscosity. The flow which results when viscous forces are overcome results in an even, streamlined flow called viscous flow. The friction force, supplied by the pressure differential, is directly proportional to the flow velocity in accordance with Darcy's law.

The relative, or normalized, flow velocity for any capillary diameter may be expressed in terms of the dimensionless *Reynolds' number* defined as:

$$\text{Re} = \frac{2\varrho Q}{\pi r \eta} = \frac{2r\bar{v}\varrho}{\eta}, \tag{3.7}$$

where Re = Reynolds' number, ϱ = fluid density, g/cm^3, r = radius of capillary, cm, η = viscosity, dyne s/cm^2, $\bar{v}$ = average linear fluid velocity, cm/s.

When the Reynolds' number exceeds approximately 2,000 in a long, straight capillary, it has been found that laminar flow begins to break down and eddies or disturbances arise in the fluid which ultimately cause the necessary pressure differential to be approximately proportional to the square of the flow rate. Therefore, the energy requirement to transfer a given quantity of fluid is greatly increased due to the turbulence introduced into the fluid. It is clear then, that if such turbulent flow occurred in wood, Darcy's law would not be obeyed because Q and ΔP would no longer be directly proportional. On the other hand, such flow is unlikely in any of the capillaries in wood except, perhaps, the largest earlywood vessels of red oak because of the high ratio of surface area to volume. It may be stated that

$\text{Re}' = 2,000$,

where Re' = critical Reynolds' number below which flow is completely viscous in a long, straight capillary.

Nonlinear flow due to kinetic energy losses where a moving fluid enters a small, short capillary is described in detail by Siau and Petty (1979) and in Sect. 3.7. It is shown by these authors that the pressure differential is proportional to the square of the volumetric flow rate, which is also approximately the same as in turbulent flow.

It has also been established that the nonlinear flow begins at Reynolds' numbers which are approximately equal to the ratio of the length to the radius of the capillary (Sect. 3.7). More exactly,

$$\text{Re}'' \approx 0.8 \, L/r \text{ for nonlinear flow}, \tag{3.8}$$

where Re'' = critical Reynolds' number below which flow is viscous and linear in a short capillary.

It is clear from Eq. (3.8) that nonlinearity can occur at very low Reynolds' numbers depending on the structure of the porous solid. Furthermore, nonlinear flow is difficult to distinguish from turbulent flow by measuring the relationship between flux and gradient because, in both cases, the pressure differential is approximately proportional to the square of the flow rate. When applied to wood structural elements where fluids enter pit openings from vessel or tracheid lumens, non-

Table 3.1. Conversion factors for permeability units

η of water at 20.2 °C = 1 cp = 0.01 poise = 0.01 dyne s/cm^2
η of air at 20 °C = 0.0181 cp = 1.81 × 10^{-4} dyne s/cm^2
1 darcy = 9.87 × 10^{-9} cm^3/cm = 9.87 × 10^{-13} m^3/m
1 cm^3/cm = 1.013 × 10^8 darcy
1 darcy = 55.3 cm^3(air)/(cm atm s) at 20 °C
1 cm^3(air)/(cm atm s) = 0.0181 darcy at 20 °C
1 cm^3(fluid)cm/dyne s = 1.013 × 10^6 cm^3(fluid)/(cm atm s)

---

linear flow could occur at Reynolds' numbers between 0.04 and 16, assuming radii from 0.005 to 2 µm, and a pit-membrane thickness of approximately 0.1 µm (Petty 1970).

Slip-flow or Knudsen diffusion is discussed in detail by Dushman (1962). It consists of molecular diffusion through a capillary due to the pressure gradient arising from the applied pressure differential. It occurs in addition to viscous flow and becomes very significant in gaseous flow when the mean free path of the gas is approximately equal to the capillary radius. Since mean free path is inversely proportional to pressure, the effect is most pronounced at low pressures, where it may constitute a large deviation from Darcy's law.

3.4 Specific Permeability

Specific permeability is equal to the product of permeability and viscosity. Thus its value is not affected by the measuring fluid and it is only a function of porous structure of the medium.

$$K = k\,\eta, \tag{3.9}$$

where K = specific permeability, darcy, cm^3/cm, or m^3/m, η = viscosity of fluid, centipoise, dyne s/cm^2, or N s/m^2, k = permeability, cm^3 (fluid)/(cm atm s), cm^3 (fluid) cm/(dyne s), or m^3 (fluid) m/N s.

The poise is a cgs unit of viscosity with dimensions of dyne s/cm^2. Since the viscosity of water is approximately 0.01 poise, corresponding to 1 centipoise, the darcy, which is the unit commonly used in the literature, is numerically equal to the water permeability. When cgs units are used throughout, specific permeability comes out in cm^3/cm, sometimes written in the literature as cm^2, although it is not entirely correct to cancel a cm measured along the specimen with one of the dimensions of the volumetric flow rate. Conversion factors for permeability units are given in Table 3.1.

Equation (3.9) is applied to liquids since the symbol k represents liquid permeability. It may also be applied to gases provided the Knudsen diffusion component of the permeability measurement is not included.

It is clear from Eq. (3.9) that specific permeability is numerically equal to the flow rate of a fluid with unit viscosity through a unit cube with a unit pressure differential between two parallel surfaces. As such, its value is a function only of the internal structure of the porous body.

Fig. 3.3. Illustrating the viscous force when two parallel surfaces move relative to each other

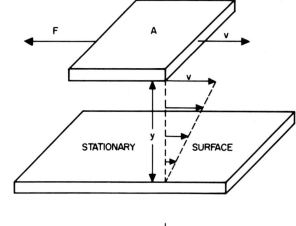

Fig. 3.4. Illustrating the equality of viscous and pressure forces on a fluid flowing in a cylindrical capillary

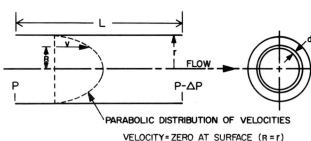

PARABOLIC DISTRIBUTION OF VELOCITIES
VELOCITY = ZERO AT SURFACE (R = r)

3.5 Poiseuille's Law of Viscous Flow

Assume two parallel surfaces as in Fig. 3.3 with the lower one stationary and the upper one moving at a constant velocity, v, through a viscous fluid. The force, F, due to the viscous reaction, is directly proportional to the viscosity, η, the surface area, A, and the velocity, v, and is inversely proportional to the distance of separation, y. Expressed mathematically,

$F = -\eta A v/y$.

The negative sign indicates that the force and velocity are oppositely directed. The quantity, v/y, is the velocity gradient or rate of shear which, in turn, is proportional to a momentum gradient. When this quantity is variable as in a cylindrical capillary or pipe, it may be expressed in derivative form as dv/dy or, more specifically, dv/dR, where R represents the variable radial distance from the axis.

Consider the viscous flow through a cylindrical capillary of radius, r, and pressure differential, ΔP, as illustrated in Fig. 3.4. The viscous force is equated to that due to the pressure differential at distance R from the center.

$-\eta A \, dv/dR = \Delta P \, \pi R^2$,

where R = radial distance at which the velocity gradient is dv/dR. From Fig. 3.4 it is clear that $A = 2\pi R L$, where L = length across which ΔP is applied.

Then, by substitution,

$$dv = -(\Delta P/2\eta L) R\, dR.$$

Since the velocity is zero at the surface, the equation may be integrated over the annular volume between the surface of the capillary of radius r and the cylindrical surface with a variable radius of R,

$$\int_0^v dv = -(\Delta P/2\eta L) \int_r^R R\, dR.$$

Performing the integration,

$$v = \frac{(r^2 - R^2)\Delta P}{4\eta L}, \tag{3.10}$$

where v = velocity at variable radial distance R, r = radius of capillary.

Equation (3.10) is parabolic in form, predicting a velocity of zero at the surface and maximum velocity at the axis. Then

$$v_{max} = \frac{r^2 \Delta P}{4\eta L}. \tag{3.11}$$

Further integration may be performed to obtain the volumetric flow rate, Q. Assume an annulus within the cylinder of thickness dR. The volumetric flow rate is then the product of the velocity and the cross-sectional area of the annulus.

$$dQ = v\, 2\pi R\, dR;$$

substituting the value of v from Eq. (3.10),

$$dQ = \frac{2\pi \Delta P (r^2 - R^2) R\, dR}{4\eta L};$$

integrating,

$$\int_0^Q dQ = \frac{2\pi \Delta P}{4\eta L} \int_0^r (r^2 - R^2) R\, dR.$$

After integration, the final form of Poiseuille's law for liquids may be written as

$$Q = \frac{N\pi r^4 \Delta P}{8\eta L}, \tag{3.12}$$

(Poiseuille's law for liquids)

where N = number of uniform circular capillaries in parallel.

When this equation is applied to short capillaries with $L/r < 100$, the *Couette correction* should be applied to account for the additional pressure drop associated with viscous losses at the ends. This is done by the substitution of a longer corrected length, L', for L. Then,

$$Q = \frac{N\pi r^4 \Delta P}{8\eta L'}, \tag{3.13}$$

(Poiseuille's law for liquids in short capillaries) where

$$L' = L + 1.2\,r. \tag{3.14}$$

The average velocity ($\bar{v}$) may be calculated by dividing Q by the cross-sectional area of the capillaries, $N\pi r^2$.

$$\bar{v} = \frac{r^2 \Delta P}{8\eta L}. \tag{3.15}$$

When Poiseuille's law is applied to gases, the expansion may be accounted for as it was in Darcy's law for gases [Eq. (3.4)].

$$Q = \frac{N\pi r^4 \Delta P \bar{P}}{8\eta L P}. \tag{3.16}$$

(Poiseuille's law for gases)

The similarity between Darcy's and Poiseuille's equations becomes clear when Eq. (3.2) is solved for Q and compared with Eq. (3.12).

$$Q = kA\Delta P/L. \tag{3.17}$$

(Darcy's law for liquids)

Similarly Eq. (3.4) for gases may be written as

$$Q = \frac{k_g A \Delta P \bar{P}}{LP}. \tag{3.18}$$

(Darcy's law for gases)

Since Q is directly proportional to the gradient, $\Delta P/L$ in Eqs. (3.12) and (3.17) for liquids and (3.16) and (3.18) for gases, it may be concluded that Darcy's law applies to the viscous flow of liquids and gases.

When Eqs. (3.12) and (3.17) are combined by the elimination of Q, it is evident that the permeability is directly proportional to the number and radii of the openings.

$$k = \frac{N\pi r^4}{A 8\eta} = \frac{n\pi r^4}{8\eta}, \tag{3.19}$$

where n = no. of capillaries per unit cross-sectional area, cm^{-2}.

Equation (3.19) may be used to calculate the permeability of a wood when its structure approximates the *parallel-uniform-circular-capillary model* in the longitudinal direction. This model is illustrated in Fig. 3.5 and is typical of diffuse-porous hardwoods with open vessels. A good example is basswood in which the vessels are relatively uniform in size. Usually a good agreement will be obtained between the measured permeability and that calculated with Eq. (3.19). This model may also be applied to an open ring-porous hardwood such as red oak because most of the flow occurs through the large earlywood vessels due to the r^4 relationship with the flow rate.

It is a consequence of Poiseuille's law that both flow rate and permeability increase very rapidly with r, for example, a doubling of r increases Q by a factor of

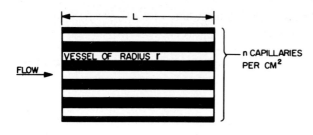

Fig. 3.5. Longitudinal flow model based upon parallel, uniform, circular capillaries

16 while a tripling of r will increase Q by a factor of 81. Since cross-sectional area is proportional to r^2, it is clear that Q increases as A^2, thus a doubling of the area will quadruple the flow rate. This may be explained by an increase in the ratio of the volume to the cylindrical area of a capillary as the radius increases, resulting in a relative decrease in the viscous friction losses per unit volume of fluid. As the radius of a capillary is increased, both v and Q increase rapidly and soon reach the point where the dimensionless velocity represented by Reynolds' number [Eq. (3.7)] exceeds the critical value of approximately 2,000 at which the flow begins the transition to turbulence.

When the uniform-parallel-circular-capillary model, characterized by Eq. (3.19), is applied to hardwoods along the fiber axis, the quantity $n\pi r^2$ becomes the fractional cross-sectional area occupied by vessels which, likewise, is equal to the volume fraction of vessels. Then Eq. (3.19) may be rewritten as:

$$k = \frac{a_v r^2}{8\eta}, \qquad (3.20)$$

where a_v = area fraction of uniform, parallel, circular capillaries.

It is clear that, if the capillaries of radius r are the only ones present in a porous body with the structure depicted in Fig. 3.5, the quantitiy a_v becomes equal to the porosity, v_a.

When Eqs. (3.19) and (3.20) are solved for specific permeability they become further simplified.

$$K = n \pi r^4/8, \qquad (3.21)$$

$$K = a_v r^2/8. \qquad (3.22)$$

It is evident from Eqs. (3.21) and (3.22) that specific permeability is determined solely by the capillary structure of the solid. In the case of the uniform-parallel-circular-capillary model, it is only a function of the porosity and the size of the openings.

3.6 Turbulent Flow

The onset of turbulence begins at $Re = 2,000$. At Reynolds' numbers well beyond this value, when the transition from viscous to turbulent flow is essentially complete, the volumetric flow velocity of a liquid in a smooth cylindrical capillary or pipe may be calculated using the following relationship (Leyton 1975).

$$Q = \frac{14.79 \, r^{19/7} \Delta P^{4/7}}{\eta^{1/7} L^{4/7} \varrho^{3/7}}. \tag{3.23}$$

(Turbulent flow of liquids)

The velocity profile in turbulent flow is essentially uniform while that for viscous flow is parabolic. In the latter case the ratio of maximum to average velocity ($v_{max}/\bar{v}$) = 2.0, while, for turbulent flow the ratio is $1/0.817 = 1.22$ (Leyton 1975).

When Eq. (3.23) is applied to gases, the expansion of the gas must be accounted for, resulting in the relationship:

$$Q = \frac{14.79 \, r^{19/7}}{\eta^{1/7} \varrho^{3/7}} \left(\frac{\Delta P \bar{P}}{PL} \right)^{4/7}, \tag{3.24}$$

(Turbulent flow of gases)

where $\bar{P}$ = average pressure, ΔP_1 = pressure differential across length, L, P = pressure at which Q is measured, ϱ = density at P.

A comparison may then be made between viscous and turbulent flow in regard to the relationships between flow rate, pressure drop, and cross-sectional area of the capillary. By referring to Eq. (3.12), it is clear that for viscous flow,

$Q \propto r^4 \quad Q \propto A^2 \quad Q \propto \Delta P \quad \text{Energy} \propto Q^2$.
(Viscous flow)

Referring to Eq. (3.23) for turbulent flow,

$Q \propto r^{2.71} \quad Q \propto A^{1.36} \quad Q \propto \Delta P^{0.57} \quad \Delta P \propto Q^{1.75} \quad \text{Energy} \propto Q^{2.75}$.
(Turbulent flow)

It is evident from the above relationships that flow rate increases much less rapidly with the size of the capillary or pipe in turbulent flow than in viscous flow, in fact the flow rate is slightly more than proportional to the area in the former case. In addition considerably greater energy is expended to sustain turbulent flow. Since energy is proportional to the product of force (ΔP) and velocity (or Q), a much greater quantity of energy is required to double Q in turbulent flow ($2^{2.75}$ or 6.7) compared with a factor of 4 in viscous flow.

3.7 Nonlinear Flow Due to Kinetic-Energy Losses at the Entrance of a Short Capillary

Tomkins (1974) presented an equation for liquid flow in which the total pressure drop across a short capillary is equal to the sum of the pressure differences due to viscous flow, as expressed by Eq. (3.12), and that due to the kinetic energy increase at the entrance of the capillary and end effects. Similar equations have been presented by Erk (1929) and Bolton and Petty (1978).

$$\Delta P = \frac{8\eta LQ}{\pi r^4} + \frac{m\varrho Q^2}{\pi^2 r^4},\qquad(3.25)$$

where m = coefficient for kinetic-energy and end-effect losses.

Mickelson (1964) found an average value of 1.19 for m from a large number of experiments. Schiller (Bolton and Petty 1978) suggested a value of 1.08. Erk (1929) explained that m must exceed unity and quotes values from 1.0 to 1.124.

Equation (3.25) may be rearranged to a more useful form by substitution of Eq. (3.7) for one of the powers of Q in the nonlinear term. Using Mickelson's value of 1.19 for m, and including the Couette correction for short capillaries, Eq. (3.25) may be written as

$$\Delta P = \frac{8QL'\eta}{\pi r^4}[1+0.074\,\mathrm{Re}(r/L')],\qquad(3.26)$$

(Liquids)

where $L' = L + 1.2\,r$.

Equation (3.26) shows that the nonlinear component of flow, due principally to kinetic energy losses at the entrance, is a function of Reynolds' number and the length-to-radius ratio of the capillary, and is independent of the radius alone. It is apparent from this equation that there is an effective critical Reynolds' number (Re'') for the onset of nonlinear flow which has a value approximately equal to the L'/r ratio of the capillary. Since most capillaries in general use have L'/r ratios much less than Re' of 2,000, it is expected that nonlinearity due to kinetic energy losses occurs at a much lower Reynolds' number than turbulence. An interesting parallel between kinetic energy and turbulence losses is that, in the former, ΔP is proportional to Q^2, and in the latter to $Q^{1.75}$ which makes the two effects indistinguishable in many flow measurements.

Equation (3.26) may be modified for gas flow by the addition of corrections for slip flow and gas expansion as proposed by Erk (1929)

$$\Delta P = \frac{8QL'\eta P}{\pi r^4 s \bar{P}}[1+0.074\,\mathrm{Re}(r/L')k],\qquad(3.27)$$

(Gases)

where s = slip-flow factor, described in Sect. 3.8, k = correction for gas expansion.

When solved for conductance, Eq. (3.27) assumes the form

$$\frac{QP}{\Delta P\bar{P}} = \frac{\pi r^4 s}{8\eta L'[1+0.074(r/L')k]}.\qquad(3.28)$$

(Adzumi equation with slip-flow and kinetic-energy corrections)

Equation (3.28) is a modified form of the Adzumi equation which will be derived in Sect. 3.8.

The value of Re'' selected by Siau and Petty (1979) was the value which resulted in a kinetic energy term which was approximately 6% of the viscous flow terms in Eqs. (3.26), (3.27), and (3.28). Therefore, it was taken as approximately 0.8 L'/r as indicated in Eq. (3.8).

The effect of nonlinear flow must be considered when short capillaries are used for viscosity or flow measurements. In addition one may speculate that nonlinear flow can occur in wood at Reynolds' numbers between 0.04 and 16 (Sect. 3.3), particularly where fluids enter a pit opening. The correction is especially important where most of the resistance to flow is in the pits. Bolton and Petty (1975) found that 75% to 86% of the resistance of air-dried Sitka spruce was in the pit openings. In regard to low critical Reynolds' numbers in porous materials in general, Scheidegger (1974) quotes several references to observed values between 0.1 and 75. Scheidegger also states that additional kinetic-energy losses can occur in curved tubes of changing diameters which could be expected in a heterogeneous substance like wood.

3.8 Knudsen Diffusion or Slip Flow

According to Fick's law, fluids diffuse from regions of higher to lower concentration with no requirement for a static pressure gradient to maintain the transport. When the diffusion occurs in a vessel whose dimensions greatly exceed the mean free path of the molecules, mutual or intergas diffusion occurs and it is controlled by intermolecular collisions. When the capillary dimensions are smaller or in the same order of the mean free path, Knudsen diffusion or slip flow occurs. In this case it is limited by collisions of molecules with the walls of the capillary.

The *mean free path* of gas molecules may be calculated from a number of similar equations. One of these, derived from the kinetic theory of gases, has the following form:

$$\lambda = \frac{2\eta}{\bar{P}} \sqrt{\frac{RT}{M_w}}, \qquad (3.29)$$

where λ = mean free path, cm, $\bar{P}$ = average pressure, dyne/cm², R = universal gas constant = 8.31×10^7 erg/mol K, T = Kelvin temperature, M_w = molecular weight = 29 g/mol for air, 28 g/mol for N_2, $\sqrt{RT/M_w} = 2.90 \times 10^4$ cm/s for air at 20 °C and 1 atm.

The mean free path of air molecules at 20 °C and 1 atmosphere is 0.1 μm. Therefore slip flow will be significant in capillary openings near this size or smaller. Slip flow is not an important factor in liquid flow, due to the relatively short mean free path of liquid molecules.

As a result of Knudsen diffusion or slip flow, the specific permeability of wood or other fine porous media is higher when measured with a gas than with a liquid. While Poiseuille's law predicts a fluid velocity of zero at the surface, this will not be the case when the slip-flow component is significant. Knudsen proposed an equation to account for this component in which the concentration gradient necessary for the diffusion is a consequence of the pressure differential causing the viscous flow. The symbols have the same significance as in Eq. (3.16), Poiseuille's equation for viscous flow of gases:

$$Q = N(0.9) \frac{4}{3} \sqrt{\frac{2\pi RT}{M_w}} \frac{r^3 \Delta P}{LP}. \qquad (3.30)$$

(Knudsen diffusion)

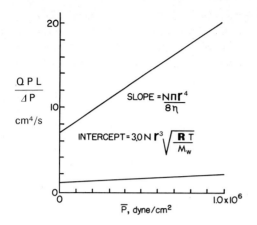

Fig. 3.6. Specific flow rate as a function of average pressure in accordance with Eq. (3.31). Slope $= N\pi r^4/8\eta$ and intercept $= 3.0\, N\, r^3\, (RT/M_w)^{0.5}$

Adzumi (1937) combined the viscous and Knudsen components of gas flow into one equation for determination of the total flow.

$$Q = \frac{N\pi r^4 \Delta P \bar{P}}{8\eta LP} + 3.0\, N \sqrt{\frac{RT}{M_w}} \frac{r^3 \Delta P}{LP}. \tag{3.31}$$

(Adzumi equation, viscous flow and Knudsen diffusion of gases)

It is clear that Q is a linear function of $\bar{P}$ because all of the other quantities are constants for a given porous body, assuming the uniform-parallel-circular-capillary model. The value of r can then be calculated from flow data of Q, Q P, Q P/ΔP, or Q P L/ΔP vs. $\bar{P}$. The Adzumi equation predicts a linear plot as revealed in Fig. 3.6 and this has been verified by several investigators (Sebastian et al. 1965; Comstock 1967; Petty 1970). It is evident from Eq. (3.31) that, when QPL/ΔP is the dependent variable,

$$\text{slope} = \frac{N\pi r^4}{8\eta},$$

$$\text{intercept} = 3.0\, Nr^3 \sqrt{\frac{RT}{M_w}},$$

then

$$r = 7.6\eta \sqrt{\frac{RT}{M_w}} \left(\frac{\text{slope}}{\text{intercept}}\right); \tag{3.32a}$$

(Adzumi plot, $\bar{P}$ expressed in dyne/cm^2)

or, alternatively,

$$r = 7.5 \times 10^{-6}\, \text{atm cm}^2/\text{dyne}\, \eta \sqrt{\frac{RT}{M_w}} \left(\frac{\text{slope}}{\text{intercept}}\right). \tag{3.32b}$$

(Adzumi plot, $\bar{P}$ expressed in atm)

Fig. 3.7. Superficial gas permeability as a function of reciprocal average pressure in accordance with Eq. (3.36)

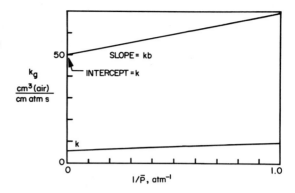

The Adzumi equation may be modified by using the superficial gas permeability as the dependent variable, with $1/\bar{P}$ as the independent variable.

$$k_g = \frac{QLP}{A\Delta P \bar{P}} = \frac{n\pi r^4}{8\eta} + \frac{3.0 r^3 \eta}{\bar{P}} \sqrt{\frac{RT}{M_w}}, \tag{3.33}$$

where pressures are expressed in dyne/cm².

This may be simplified further by the substitution of mean free path [Eq. (3.29)] and by the substitution of k for the viscous term [Eq. (3.19)]. This results in the Klinkenberg equation

$$k_g = k\left(1 + \frac{3.8\lambda}{r}\right) = ks, \tag{3.34}$$

(Klinkenberg equation)

where k = permeability in absence of slip flow.

From Eq. (3.31) it may be shown that the slip-flow factor in the Adzumi equation is equal to

$$s = 1 + \frac{7.6\eta}{\bar{P}r}\sqrt{\frac{RT}{M_w}}, \tag{3.35}$$

where $\bar{P}$ is expressed in dyne/cm².

The value of k in Eq. (3.34) can be obtained by the extrapolation of k_g to zero reciprocal average pressure. It is clear from this equation that the Knudsen diffusion term increases with the ratio of mean free path to capillary radius. Since λ increases with decreased average pressure, Knudsen diffusion and k_g increase with decreased average pressure. The viscous and slip-flow terms are equal at the average pressure at which r is 3.8 times the mean free path of the gas.

Equation (3.34) may also be written as

$$k_g = k(1 + b/\bar{P}), \tag{3.36}$$

where k = intercept, $b = \dfrac{3.8\lambda \bar{P}}{r} = \dfrac{7.6\eta}{r}\sqrt{\dfrac{RT}{M_w}}.$

Typical plots based upon data of permeability vs. reciprocal pressure are revealed in Fig. 3.7. The intercept is k and the slope is k b. As in the Adzumi equation, r may be calculated from the slope and intercept.

$$r = 7.6\eta \sqrt{\frac{RT}{M_w}\left(\frac{\text{intercept}}{\text{slope}}\right)}; \qquad (3.37\text{a})$$

(Klinkenberg plot, $\bar{P}$ expressed in dyne/cm^2)

or,

$$r = 7.5 \times 10^{-6} \text{ atm cm}^2/\text{dyne } \eta \sqrt{\frac{RT}{M_w}\left(\frac{\text{intercept}}{\text{slope}}\right)}. \qquad (3.37\text{b})$$

(Klinkenberg plot, $\bar{P}$ expressed in atm)

Equations (3.37a) and (3.37b) are identical with Eqs. (3.32a) and (3.32b) with the slope and intercept reversed in correspondence with the change of the independent variable from $\bar{P}$ to $1/\bar{P}$.

If air at 20 °C is used as the medium and the pressure is expressed in atmospheres, Eqs. (3.32b) and (3.37b) simplify to:

$$r = 0.40 \times 10^{-4} \text{ atm cm}^2/\text{dyne}\left(\frac{\text{slope}}{\text{intercept}}\right), \qquad (3.38)$$

(Adzumi plot with $\bar{P}$ expressed in atm)

and

$$r = 0.40 \times 10^{-4} \text{ atm cm}^2/\text{dyne}\left(\frac{\text{intercept}}{\text{slope}}\right). \qquad (3.39)$$

(Klinkenberg plot with $1/\bar{P}$ expressed in atm^{-1})

3.9 Corrections for Short Capillaries

There are two corrections which may be applied to the Adzumi and Klinkenberg equations for the determination of the radius of short capillaries. This may be necessary in the case of pit openings which have lengths in the order of 0.1 micrometer (Sebastian et al. 1965; Petty 1970).

It will be recalled that the Couette correction is applied to the Poiseuille equation to account for end resistance in a short capillary. This is accomplished by increasing the effective length from L to $L' = L + 1.2$ r. The factor is in the denominator of Eq. (3.13) resulting in a decreased Q. A calculated radius from an uncorrected equation will therefore be low and it may be corrected by multiplying the result by the factor $(1 + 1.2 \text{ r}/L)$.

The *Clausing factor* is applied to the slip-flow term to account for decreased Knudsen diffusion in a short capillary. The factor, K_c, is substituted for 8 r/3 L. Dushman (1962) gives approximate Clausing factors as:

$K_c = 1/(1 + 0.5 \text{ L}/r)$ $K_c = 1/(1 + 0.375 \text{ L}/r)$
for $L/r \leq 1.5$ for $L/r > 1.5$

The factor applied to a calculated radius using the uncorrected equation is then $K_c \div 8\, r/3\, L$ or $3/8\, (K_c\, L/r)$. Since the factor is always less than one and is applied to the slip-flow term which appears in the denominator of the calculation for r, it has the effect of reducing its value below that of the uncorrected equation.

The overall correction factor for r is then:

$$\frac{L/r+1.2}{1.33\, L/r+2.67} \quad \text{for} \quad L/r \leq 1.5 \tag{3.40a}$$

and

$$\frac{L/r+1.2}{L/r+2.67} \quad \text{for} \quad L/r > 1.5. \tag{3.40b}$$

(Combined Couette correction and Clausing factor to be applied to calculation of r)

It is fortuitous that the Couette correction and the Clausing factor tend to cancel each other when applied to the Adzumi and Klinkenberg equations for the calculation of r. It is clear from Eqs. (3.40a) and (3.40b) that the limits of the overall correction factor extend from 0.45 for very low L/r to 1.0 for a high L/r. Therefore, under extreme conditions, an uncorrected radius could be high by a factor of 1/0.45 or 2.2.

3.10 Permeability Models Applicable to Wood

3.10.1 Simple Parallel Capillary Model

The simplest model, and that upon which the Poiseuille equation is based, is the uniform-parallel-circular-capillary model depicted in Fig. 3.5. As previously stated, this may be applicable to the open vessels of diffuse porous hardwoods in the longitudinal fiber direction.

The radius r may be calculated from the slope and intercept of the Klinkenberg plot using Eq. (3.37a) or (3.37b). It is then possible to calculate n from the intercept permeability using Eq. (3.19) if the pressure is expressed in dyne/cm². If the pressure is expressed in atmospheres, a conversion factor must be used:

$$k = \frac{n\pi r^4}{8\eta} \times 1.013 \times 10^6 \, \frac{\text{dyne}}{\text{cm}^2 \, \text{atm}}, \tag{3.41}$$

where k is expressed in cm³ (fluid)/(cm atm s).

Smith and Lee (1958) have used this model to compare theoretical longitudinal air permeabilities of hardwoods, calculated from measured values of n and r, with experimental values. A reasonable agreement between the two was found for the several species of hardwoods measured.

3.10.2 Petty Model for Conductances in Series

Plots of permeability vs. $1/\bar{P}$ or specific flow rate vs. $\bar{P}$, for wood specimens are frequently curvilinear as revealed in Fig. 3.8 (Petty 1970; Siau et al. 1981). This has

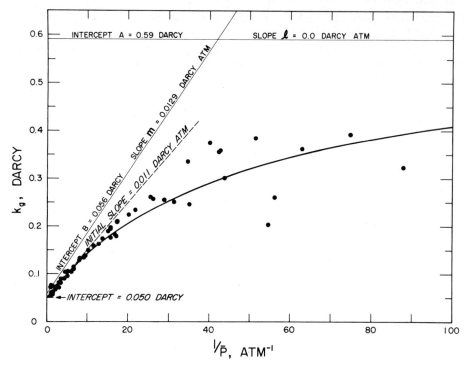

Fig. 3.8. Longitudinal steady-state permeability vs. reciprocal pressure for a specimen of *Flaxinus americana*. The *curved line* is a least-squares fit calculated from a gradient-search program. The component linear Klinkenberg plots for the series elements are shown

been attributed by Petty to the presence of high and low conductances acting in series, for example, the tracheids and pit openings in softwoods, and vessels and intervessel pits in hardwoods. For example, the measured conductance may be related to two or more components in the following way:

$$\frac{1}{g} = \frac{1}{g_t} + \frac{1}{g_p}, \tag{3.48}$$

where g = conductance of two resistances in series, g_t = conductance of tracheid portion, g_p = conductance of portion through pit openings.

The conductance may be expressed in terms of longitudinal gas permeability by use of the relationship,

$$g = k_{gL} A/L. \tag{3.49}$$

Since A is common to both portions, it divides out. Substituting Eq. (3.49) into Eq. (3.48),

$$\frac{L}{k_{gL}} = \frac{L_{Lt}}{k'_{Lt}} + \frac{L_{Lp}}{k'_{Lp}}, \tag{3.50}$$

where L = specimen length, L_{Lt} = total length of tracheid portion, L_{Lp} = total length of pit-opening portion, such that $L = L_{Lt} + L_{Lp}$, k'_{Lt} = true superficial gas permeability of tracheid portion, k'_{Lp} = true superficial gas permeability of the pit-opening portion.

Since the pit membrane is very thin (≈ 0.1 μm), it is apparent that the total of the membrane thickness, $L_{Lp} \ll L_{Lt}$. Therefore L_{Lp} is negligible in comparison with L_{Lt} and $L_{Lt} \approx L$. Then, writing Eq. (3.50) in terms of permeabilities,

$$\frac{1}{k_{gL}} = \frac{1}{k'_{Lt}} + \frac{L_{Lp}}{k'_{Lp}L}. \tag{3.51}$$

The term (k'_{Lt}) of Eq. (3.51) is represented in the Petty model by the expression $(A + l/\overline{P})$ where A is the intercept permeability and l is the slope of the function of k'_{Lt} vs. $1/\overline{P}$ in accordance with the Klinkenberg equation. The term ($k'_{Lp} \cdot L/L_{Lp}$) represents the conductance of the pit-opening portion and is written in the form of $(B + m/\overline{P})$ where B is the intercept conductance and m is the slope of the straight line function of conductance vs. $1/\overline{P}$. Therefore, the equation for superficial gas permeability in accordance with the Petty model of two conductances in series may be written as

$$k_{gL} = \frac{(A + l/\overline{P})(B + m/\overline{P})}{(l + m)/\overline{P} + A + B}. \tag{3.52}$$

By use of a nonlinear curve-fitting computer program, the experimental data of k_g vs. $1/\overline{P}$ may be used to generate values of the slopes and intercepts of the two components. The slopes and intercepts may then be used to calculate the respective radii from Eqs. (3.37a) or (3.37b). The correction factor for short capillaries can be applied as described in the previous section. Petty (1970) has used this to characterize the structure of the wood of *Picea sitchensis*. Equation (3.52) may also be extended to apply to three series components (Bolton and Petty 1975). This model may be employed to determine the relative resistances of the components along with the number and size of openings, as will be described later.

A nonlinear relationship between k_{gL} and $1/\overline{P}$ may also be caused by nonlinear flow due to kinetic-energy losses or to turbulence (Sect. 3.7). Either of these is easily detected by an increase in the calculated permeability when the flow rate is decreased at a given average pressure. When the flow is linear and obeys Eq. (3.52), the permeability is independent of flow rate at a constant average pressure. Therefore, when nonlinearity or turbulence is present there will be a large scatter in the k_{gL} vs. $1/\overline{P}$ data points unless they are all taken at the same Reynolds' number.

3.10.3 Comstock Model for Softwoods

The flow of fluids through softwoods is essentially through the tracheids which are interconnected by bordered pit pairs. The pit openings are very small in diameter compared with the lumens and are therefore assumed to provide all the flow resistance. Therefore it is the number and condition of the pit openings which determines the permeability. The model proposed by Comstock (1970) further assumes that all the pits are on the radial surfaces at the tapered ends, as illustrated in Fig. 3.9. All pit openings are assumed equal in size. Figure 3.9 shows four pit pairs

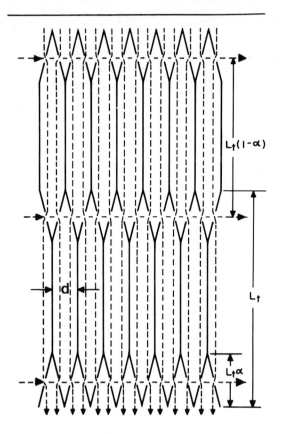

Fig. 3.9. Softwood flow model according to Comstock (1970). Tangential section showing pits on the radial surfaces of the tapered ends of the tracheids

open to each tracheid which are shared by two tracheids resulting in two pit pairs per tracheid. It is evident that both longitudinal and tangential permeabilities are increased by the relative number of bordered pit pairs in parallel per unit cross-sectional area perpendicular to the flow direction and are decreased by the number in series per unit length in the flow direction.

In order to derive the ratio of longitudinal to transverse (tangential) permeabilities, it is only necessary to set up proportions for both the longitudinal and tangential values.

In the longitudinal direction there are two parallel paths per tracheid, or $2/(2\,r_t)^2$, or $1/(2\,r_t^2)$ per cm^2 of cross-sectional area, where r_t = radius of tracheid. The number of pit openings traversed per cm of length is $1/[L_t(1-\alpha)]$, where L_t = length of tracheid and α = fraction of overlap. Then,

$$k_L \propto \frac{1/2r_t^2}{1/L_t(1-\alpha)} = \frac{L_t(1-\alpha)}{2r_t^2}.$$

In the tangential direction there is one parallel path per tracheid, or $1/[2\,r_t L_t (1-\alpha)]$ per cm^2 of area. The number of pit openings traversed per cm of length is $1/r_t$. Then

$$k_T \propto 1/[2\,L_t(1-\alpha)].$$

Therefore,
$$\frac{k_L}{k_T} = \frac{L_t^2(1-\alpha)^2}{r_t^2}. \tag{3.53}$$

Assuming a length-to-diameter ratio of 100 for the tracheids, $L_t/r_t = 200$, and Eq. (3.53) may be simplified to:
$$\frac{k_L}{k_T} = 40{,}000\,(1-\alpha)^2. \tag{3.54}$$

Equation (3.54) reveals that the ratio k_L/k_T can vary between 10,000 and 40,000 as α decreases from the maximum value of 0.5 to zero. This is in reasonable agreement with experimental results (Comstock 1970).

It is conceivable that this model may apply to hardwoods in which the vessels are closed by tyloses, since the flow would then be between fiber cells interconnected by pit pairs. As for hardwoods with open vessels, these would be expected to have extremely high ratios of k_L/k_T, possibly as much as $10^6:1$ because of the highly conductive vessels.

3.10.4 Characterization of Wood Structure from Permeability Measurements

The resolution of k_{gL} vs $1/\bar{P}$ (or $QP/\Delta P$ vs $\bar{P}$) into viscous and slip-flow components makes it possible to calculate radii and numbers of openings for each of the components. In the case of longitudinal measurements using the Petty model, the radius of the large component, r_1, may be calculated from the slope and intercept as described previously. Equation (3.36) may be adapted to this model for these calculations. The term, k, is replaced by the intercept A. Although A is a conductance it may be assumed equal to the permeability of this component because the length of the flow path through the pit openings is very small compared with that through the lumens of the tracheids or vessels and may therefore be neglected. The radius may be determined as

$$r_1 = 0.40 \times 10^{-4}\ (\text{atm cm}^2/\text{dyne})\ (A/I), \tag{3.55}$$

where A/I is in atm^{-1}, r_1 = radius of tracheid or vessel lumens, cm.

Referring to Eqs. (3.51) and (3.52), it is clear that the term A is the intercept permeability for the tracheid path, k_{Lt}. Then Eq. (3.19) may be modified for the calculation of the number of conductive tracheids per cm² of cross-sectional area

$$A = k_{Lt} = \frac{n_{tc}\pi r_1^4}{8\eta}, \tag{3.56}$$

where n_{tc} = number of conductive tracheids per cm² of cross-sectional area.

In the case of the pit-opening portion, the radius of the pit openings may be calculated as

$$r_2 = 0.40 \times 10^{-4}(\text{atm cm}^2/\text{dyne})\ (B/m). \tag{3.57}$$

In the case of pit openings, it must be recognized that the conductance B must be multiplied by a length factor to convert it to a permeability to permit calculation of the total number of conductive pit openings per cm² (n_p). Let L_p be the thickness or length of path through the pit openings. The longitudinal path length between pit pairs at opposite ends of the tracheid is then $L_t(1-\alpha)$ according to the Comstock model. Therefore a length factor of $L_p/[L_t(1-\alpha)]$ is applied to the calculated intercept conductance of the pit-opening path (B) to obtain its true permeability. In the calculation of the number of openings, the term B is the intercept conductance which may be written as

$$B = \frac{k_{Lp}L}{L_{Lp}}, \tag{3.58}$$

where k_{Lp} = true intercept permeability of the pit openings, L_{Lp} = total length of the pit membranes traversed in the specimen.

The ratio L_{Lp}/L may be represented more simply by the ratio of L_p, the thickness of one membrane to $L_t(1-\alpha)$, the length of the path between two pit openings in one tracheid according to the Comstock model (Fig. 3.9).

Then,

$$k_{Lp} = \frac{BL_p}{L_t(1-\alpha)} = \frac{n_p \pi r_2^4}{8\eta}, \tag{3.59}$$

where n_p = number of conductive pit openings per cm² of cross section.

The number of conductive pit openings per conductive tracheid (n_{pt}) may then be calculated as:

$$n_{pt} = n_p/n_{tc}. \tag{3.60}$$

The total number of tracheids per cm² may be calculated from the radius of the tracheids, r_t.

$$n_t = 1/(4 r_t^2) \tag{3.61}$$

and,

fraction of conductive tracheids = n_{tc}/n_t. (3.62)

Petty (1970), using a staining technique, found 2.7% to 7.8% of the tracheids of *Picea sitchensis* were conducting in the longitudinal direction. Most of these were located in the latewood. This result was in good agreement with the value of n_{tc} calculated from flow measurements. He also found an average of 250 openings for each conducting tracheid (n_{pt}), which he interpreted as openings in the margo of the pit membrane. This was in reasonable agreement with results determined from electron micrographs.

In the *tangential* direction Petty (1970) has found a negligible resistance in the lumen path of *Picea sitchensis* because a linear relationship was found between $Q/\Delta P$ and $\bar{P}$. The radius of the pit openings could then be calculated as described in Sect. 3.8. Since the Klinkenberg equation would apply, the number of conductive pit openings per cm² in the tangential direction may be calculated from the true permeability which is determined from the measured tangential intercept permeability by the application of a length factor. In tangential flow, the length factor is L_p/r_t because one pit pair is traversed for a path length equal to the radius. Then the number of pit openings per cm² for tangential flow may be calculated from the relationship

$$\frac{k_T L_p}{r_t} = \frac{n_{pT} \pi r_2^4}{8\eta}, \tag{3.65}$$

where n_{pT} = number of pit openings per cm^2 calculated from tangential flow, k_T = tangential intercept permeability.

The thickness of the pit membrane, L_p, has been measured by Sebastian et al. (1965) as 0.15 μm in the wood of *Picea glauca* and Petty (1970) has used a value of 0.1 μm. Either of these values is reasonable if no direct measurements are available. Values of L_t, α, and r_t may be measured microscopically.

Other aspects of the use of permeability and capillary phenomena to characterize wood structure are discussed in detail by Siau et al. (1981). In this publication it is described how the relative inhomogeneity of a porous body may be evaluated by calculations of the mean effective radius from a Klinkenberg plot or by use of the Petty model, by measuring the bubble point, by microscopic observation, and by calculation from gas flow using the parallel capillary model. If such determinations were done with a porous solid consisting of uniform, circular, parallel capillaries, all the calculated radius values would be equal. It follows from this that the more nonhomogeneous the porous structure, the greater the deviation will be between values of r determined by the different methods. This emphasizes one of the principal benefits of research efforts in the field of wood-permeability measurement, namely, the ability to use flow measurements to describe anatomical structure by the determination of the sizes of pit openings, tracheids, and vessels, and the number of openings by which the various cell types are interconnected.

3.11 Measurement of Liquid Permeability

The details of the techniques for the measurement of the liquid permeability of wood will not be described since they have been elucidated by Kelso et al. (1963). The flow of liquids through wood is greatly complicated by capillary forces due to entrapped air which may greatly exceed the forces of viscous resistance which determine the permeability. Thus it is necessary to deaereate the liquid to as great a degree as possible. In addition, liquids contain particulate matter which can clog the minute pit openings and these must be removed by microfiltration. Liquid permeability has also been measured successfully by Kininmonth (1971) and by Comstock (1965, 1967). Comstock (1967) determined the permeability of the sapwood and heartwood of *Picea glauca* with nonswelling liquids and nitrogen gas and found that specific permeabilities were equal when the viscosity difference and the molecular slip flow of the gas were accounted for.

3.12 Measurement of Gas Permeability

The procedure for the measurement of gas permeability is much simpler due to the elimination of the problems associated with capillary forces. Also, the principal deviations from Darcy's law in gas flow: gas expansion and Knudsen diffusion, can

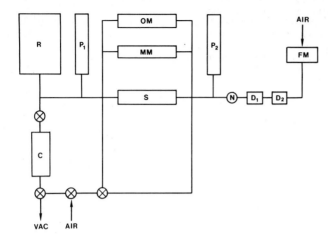

Fig. 3.10. Schematic diagram of Petty apparatus; C Cartesian manostat; R vacuum reservoir; P_1, P_2 open mercury manometers for vacuum and high-pressure ends of specimen; MM differential mercury manometer; OM differential oil manometer; S test specimen; N needle valve; DS silica-gel desiccant; DP P_2O_5 desiccant; FM flow measurement apparatus consisting of either calibrated burets, rotameters, or electronic flowmeters

Fig. 3.11. Steady-state permeability apparatus designed by Petty. Symbols as in Fig. 3.10

be accounted for. In fact, the latter is an advantage because it makes it possible to calculate radii and to separate high and low permeability components in series. Gas measurements are most conveniently made with air, but precautions are necessary to remove moisture and particulates by an adequate desiccant and microfiltration.

An excellent apparatus for permeability measurement, illustrated in Fig. 3.11 was designed by Petty and Preston (1969). A schematic diagram is presented in Fig. 3.10. Three manometers are shown although only two are necessary to determine ΔP and $\bar{P}$. The apparatus in Petty's laboratory is simplified by using a closed manometer on the vacuum side, making it unnecessary to use the barometric pressure in the calculation of $\bar{P}$.

With an apparatus of the type depicted in the schematic, the permeability may be calculated using the following equation:

$$k_g = \frac{76(\text{cm Hg/atm})QLP_a}{A\Delta P(P_a - \Delta P_1 + \Delta P/2)}, \tag{3.66}$$

where k_g = superficial gas permeability, cm^3(gas)/(cm atm s), Q = flow rate, cm^3/s, L = length of specimen, cm, A = area of specimen, cm^2, P_a = atmospheric pressure, cm Hg, ΔP = pressure differential across specimen, cm Hg, ΔP_1 = open manometer reading on the vacuum side, cm Hg.

The average pressure may also be calculated as $(P_a - \Delta P_2 - \Delta P/2)$, where ΔP_2 = open manometer reading on the high-pressure side. Note that the flow rate is measured at atmospheric pressure, accounting for the term (P_a) in the numerator. Using this apparatus it is possible to measure permeabilities at average pressures from 1 cm Hg to atmospheric pressure. It may be modified to read permeabilities above atmospheric pressure. An oil manometer is recommended for improved accuracy in reading low values of ΔP. The apparatus is made from glass tubing with interchangeable ground-glass joints to permit easy disassembly and modification, and to minimize pressure losses between the manometers and the test specimen. Flow rate may be measured with rotameters or electronic flowmeters, the latter permitting the measurement of lower flow rates, being available with a full scale of 5 cm^3 per minute and a minimum readable flow of 0.2 cm^3 per minute. Petty has used calibrated buret tubes with a moving soap bubble and a stop watch with high accuracy.

It is also possible to use rising water in a glass displacement tube as a means of measuring gas flow rate. The method is described by Siau (1971). It is a simple apparatus to construct but has the disadvantage of being an unsteady-state method due to the changing pressure as the water rises in the tube. A falling-water displacement method, particularly well suited to very permeable specimens, is also described by Siau.

3.13 The Effect of Drying on Wood Permeability

The kiln- or air-drying of wood results in high capillary forces at the surface of the free water, as explained in detail in Chapter 4.6. These forces can cause aspiration in pits of woods of the Pinaceae family which generally have an impermeable torus. Other softwood genera, including *Taxodium*, *Juniperus*, *Sequoia*, and *Thuja*, in addition to all of the hardwoods, do not have pit tori and therefore their permeabilities would not be affected in this manner. Phillips (1933) has found by microscopic observation that the incidence of pit aspiration, particularly in earlywood, increases rapidly as the wood moisture approaches the fiber saturation point. Latewood is much more resistant to this effect particularly in softwoods with an abrupt transition between earlywood and latewood as in *Pseudotsuga menziesii*. This is attributed to the smaller diameter and thicker membranes in the latewood resulting in much greater mechanical rigidity. Bolton and Petty (1978) have calculated the pressure required to deflect the pit membranes of the wood of *Pinus sylvestris* and obtained values of 5.6 atm in earlywood and 680 atm in latewood. Therefore, the latewood, despite its smaller and fewer pits, usually has a much higher permeability when dry, although the reverse would be expected in the living tree. The tendency toward pit aspiration would be expected to increase with increased drying temperatures due to the decrease in strength with temperature (Comstock 1968).

Bramhall and Wilson (1971) observed decreases in permeability resulting from air-drying of sapwood earlywood of *Pseudotsuga menziesii*. The heartwood and latewood were essentially unaffected. Comstock and Côté (1968) observed greatly reduced permeability in sapwood specimens of *Pinus resinosa* and *Tsuga canadensis* as a result of air-drying at temperatures from -18 °C to 140 °C, both with slow drying with no circulation over P_2O_5 in a desiccator and rapid drying with air circulation in an oven. Permeability of dried material decreased with an increase in both temperature and drying rate. Permeabilities of dried specimens were approximately 5% of green material when dried at -18 °C decreasing to approximately 1% at 140 °C.

In summary, a decrease in permeability due to pit aspiration during air-drying occurs in the Pinaceae family which includes a large portion of woods treated with preservative solutions. The greatest effect occurs in sapwood earlywood, which explains the generally superior treatability of the latewood of softwoods.

3.14 Treatments to Increase Permeability

Pit aspiration during drying may be avoided or reduced by modification of the drying procedure for the elimination or reduction of surface tension forces. (See Chap. 4.6.) Two methods have been used to accomplish this: *freeze-drying* and solvent-exchange drying. In the former, the green wood is frozen and evacuated under an absolute pressure of less than 0.46 cm Hg which corresponds to the triple point of water below which ice sublimes directly to vapor with no liquid phase. Surface-tension forces occurring at the water–air interface where air-drying occurs are thereby eliminated.

Surface tension forces can be significantly reduced by *solvent-exchange drying* in which the sap or free water is exchanged with liquids having a much lower surface tension. Bramhall and Wilson (1971) employed alcohol-benzene extraction to accomplish this. It is also possible to bring about the exchange by diffusion at room temperature by successive immersion of the wood specimens in a miscible solvent. Several such exchanges may be required and therefore the process is time-consuming. In Bramhall and Wilson's investigation, air- and oven-drying were compared with solvent-exchange and freeze-drying of *Pseudotsuga menziesii* wood. The drying had essentially no effect on any latewood specimens, which had permeabilities of approximately 0.2 darcy. Also, the permeability of heartwood earlywood was relatively low (0.002 to 0.2 darcy) and unaffected by the drying method because of the large fraction of aspirated and encrusted pits in the undried material. The solvent-exchange and freeze-dried earlywood sapwood had higher permeability than the latewood (approximately 0.5 to 2 darcys in interior type and 4 to 8 darcys in coastal type). This result would be expected because of the larger number and larger diameter of pits in earlywood. In coastal-type wood the air-dried material had a permeability of 1% to 10% of that dried by low-surface-tension methods, and in interior type the range was from 5% to 25%.

Petty (1978a) compared the treatability of the sapwood of *Picea sitchensis* which was solvent-exchange dried with methanol or acetone with similar air-dried specimens and found much higher rates of absorption of petroleum distillate in the former. The difference was attributed to a reduction of pit aspiration.

While solvent-exchange and freeze-drying have been used successfully to reduce pit aspiration and therefore increase permeability, they are essentially applicable to small specimens as a powerful research tool to elucidate the reasons for reduced permeability during drying; they are impractical on a commercial scale.

When the decreased permeability is due to causes other than pit aspiration, chemical or microbiological means may be used to increase permeability. Lantican et al. (1965) treated the wood of *Thuja plicata*, conditioned to 20% moisture content, with an ozone–oxygen mixture consisting of 3.2% ozone. The result was an increase of approximately 100-fold in the permeability, along with an increase in hygroscopicity and a loss in dry weight. The treatment attacks the secondary wall, resulting in delignification and in degradation of incrustations in the pit membranes. Nicholas and Thomas (1968a) treated the sapwood of *Pinus taeda* with the enzymes pectinase and hemicellulase. Electron micrographs disclosed enzymatic degradation of the torus with the greatest effect from the pectinase. The pectinase treatment thus increased the permeability, especially after presteaming at atmospheric pressure for 4 h. The use of bacterial and enzyme attack of pit membranes as a result of ponding has been studied by Unligil (1972). He found that a ponding period of 5 weeks in a natural lake was sufficient to promote full sapwood penetration of *Picea glauca* wood. Creosote retention of the ponded material was 155% greater than that of bolts which were stored indoors. A reduction of modulus of rupture of approximately 4% in sapwood occurred after 9 weeks of ponding. Dunleavy and McQuire (1970) and Dunleavy et al. (1973) conducted extensive investigations of commercially ponded bolts of *Picea sitchensis* and *Picea excelsa* which are normally refractory to preservative treatment. It was found that the treatability of the sapwood was improved very significantly in the tangential, radial, and longitudinal directions, while that of the heartwood was essentially unaffected. The tori and pit membranes were extensively degraded by bacterial attack, while preliminary tests revealed no significant loss of mechanical strength. The increase in radial permeability was attributed to degradation of the cross-field pit membranes in the ray parenchyma cells. Complete penetration of sapwood was found along the entire length of poles 9 m long. Ponding for 6 to 8 weeks followed by drying for 6 weeks therefore appears to be a commercially feasible process for improvement of the sapwood treatability of these refractory woods.

Nicholas and Thomas (1968b) describe various investigations of the effect of steaming on the permeability of wood. They cite a study by Benvenuti (1963 unpublished results) in which a 30-fold increase in permeability after drying of *Pinus taeda* sapwood was obtained after presteaming at atmospheric pressure for 4 h. Transmission electron micrographs of pit membranes of the steamed wood indicated hydrolysis of the membranes which would weaken both the membrane and the bond between the torus and the border, resulting in deaspiration of the pits.

While such a process may be applied relatively inexpensively to lumber, great care must be exercised to prevent mechanical degradation due to hydrolysis of the constituents of the cell wall.

Thomas and Kringstad (1971) were able to deaspirate pits by soaking specimens of *Pinus taeda* wood in water for periods of one and four weeks. When air-dried wood was stored for 3 months, then soaked for 1 week, solvent-exchanged with pentane, and dried, all of the pits were deaspirated. On the other hand, a 13-month storage period resulted in deaspiration of 40% of the pits by the same treatment, indicating that additional bonding between the torus and border results from the longer storage period.

3.15 The Effect of Moisture Content on Permeability

It is obvious that the free water must be removed from wood before it may be impregnated with a preservative solution. Wood above the fiber saturation point would be expected to have a very low permeability because high capillary pressures must be overcome to force air bubbles through the minute pit openings as will be discussed in Chap. 4. On the other hand, if the wood were completely saturated with a nonpolar liquid, there would be no air blockage, and the specific liquid permeability would be equal to that for gases extrapolated to zero reciprocal pressure (Comstock 1967).

Comstock (1968) has investigated the effect of moisture content below the fiber saturation point on the permeability of several softwoods and hardwoods. Generally, the permeability of softwoods increases as moisture content decreases, but the difference is not highly significant in terms of its effect on liquid impregnation. Typically, the increases are two- to threefold for a moisture content decrease from 24% to 6%. Comstock attributed the change to shrinkage of the microfibrillar strands in the pit margo. Several hardwoods exhibited the opposite effect with increasing longitudinal permeability with increased moisture content. This could be due to an increase in the fractional volume of the vessels.

3.16 The Influence of Specimen Length on Permeability

The deviations from Darcy's law due to gas expansion, molecular slip flow, end effects, nonlinear flow, and turbulent flow have been discussed. Permeability has also been found to vary with specimen length under certain conditions. Sebastian et al. (1965) found that the longitudinal air permeability of *Picea glauca* wood decreased as the length was increased from 0.32 cm to 2 cm. Bramhall (1971) made similar observations with the wood of *Pseudotsuga menziesii* with lengths between 0.5 and 3.5 cm. The effect was greater in specimens of lower permeability. Siau (1972) conducted a similar investigation with the same wood in which 30-cm-long

Fig. 3.12. The relationship between the logarithm of the permeability and the length of end-matched specimens of *Pseudotsuga menziesii* (Douglas-fir) and *Pinus taeda* (Loblolly pine). *LP* loblolly pine; *DF (CS)*, Pacific-coast sapwood of Douglas-fir; *DF (IS)*, intermountain sapwood of Douglas-fir; *DF (IH)*, intermountain heartwood of Douglas-fir. (Siau 1972)

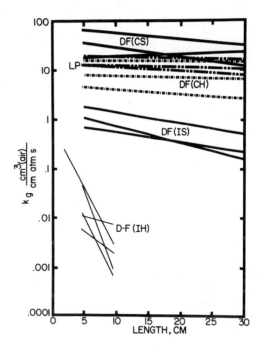

specimens were cut to successively shorter lengths and the permeabilities measured. The results are shown in Fig. 3.12 where it is evident that the decrease in the logarithm of the permeability with length is small at relatively high permeabilities. Both Bramhall and Siau detected a negligible effect at permeabilities exceeding approximately 2 darcys. On the other hand, Kumar (1981) observed no significant change in woods with permeabilities of 0.01 to 22.5 darcy when the length was decreased from 5 cm to 2.5 cm.

Bramhall (1971) explained the decrease in permeability with length by a model in which there is an exponential decrease in the effective area for conduction with increasing length due to a progressive closing off of parallel flow paths within the wood. This can be expressed mathematically as

$$\text{Effective conductive area} = A\, e^{-bL}, \tag{3.67}$$

where A = cross-sectional area, L = length, b = positive exponential coefficient obtained from the slope of the plot of ln K vs L.

Using the notation for specific permeability, the apparent permeability of a specimen with length L is calculated from Darcy's law as

$$K = \frac{VLP\eta}{tA\Delta P\bar{P}}. \tag{3.68}$$

The permeability at zero length may then be calculated by substitution of Eq. (3.67) into Eq. (3.68).

$$K_0 = \frac{VLP\eta}{tAe^{-bL}\Delta P\bar{P}}; \tag{3.69}$$

then, by substitution

$$K = K_0 e^{-bL}; \qquad (3.70)$$

or, in logarithmic form,

$$\ln K = \ln K_0 - bL. \qquad (3.71)$$

Equation (3.71) reveals that the intercept of a plot of ln K vs L is the permeability at zero length while -b is the slope of the plot. Actually, K_0 has little physical meaning because, if a specimen is cut shorter than a cell length, the permeability will rise to a very high value corresponding to capillaries with the diameters of the tracheids because the resistance of the pit openings will be eliminated.

Perng (1980a, b) has investigated the influence of length on permeability in softwoods and hardwoods of high and low permeability under conditions of constant pressure gradient and constant pressure differential. It was found that permeability decreases with an increase in length in woods with permeabilities of less than 2 darcys with a constant pressure gradient in agreement with the results of Bramhall (1971) and Siau (1972). In hardwoods with permeabilities between 2 and 10 darcys the permeability was observed to increase sharply with length up to approximately 10 cm and then to level off. This increase could be caused by nonlinear flow due to kinetic-energy losses at the entrance of the vessels, as discussed in Sect. 3.7. This effect would increase as the length to diameter ratio of the vessels decreases. Also, an increase in permeability with length could be a result of end effects, particularly the condition of the end surface, because it was found that a smoother surface resulted in increased permeability.

3.17 Permeability of the Cell Wall

All the models described in Sect. 3.10 were based upon the assumption of zero permeability of the cell wall. It is known, however, that polar liquids such as water can diffuse through the cell wall. Palin and Petty (1981) describe an experimental procedure used for the measurement of the permeability of the cell wall of *Picea abies*. Initially the voids of the wood were filled with paraffin wax which did not penetrate the cell wall. Due to the extremely low permeability, it was not possible to use a static pressure gradient for the determination. Instead, a gradient of osmotic pressure was applied using solutions of polyethylene glycol and dextran of sufficiently high molecular weight to prevent the solute from entering the cell wall. The average values obtained were 68×10^{-17} cm^3/cm in the longitudinal direction, 7×10^{-17} cm^3/cm in the radial direction, and 4×10^{-17} cm^3/cm in the tangential direction. These correspond to 7×10^{-8}, 0.7×10^{-8}, and 0.4×10^{-8} darcy respectively. Because these values are several orders of magnitude below any measured permeabilities of wood, the assumption of zero cell-wall permeability in the models is justified.

3.18 Zones of Widely Differing Permeabilities in Wood

Nondarcian behavior of fluids in wood may result from the presence of zones with vastly different permeabilities which represent a significant departure from the uniform-parallel-capillary model. This has been discussed in Sect. 3.10, where an example is the presence of high and low permeabilities in series elucidated by the Petty model. In this case the curvilinear relationship between permeability and reciprocal pressure represents a departure from Darcy's law even though each conductance taken individually behaves in accordance with the Klinkenberg equation.

Parallel zones with different flow properties occur in the red oaks which have very large, open earlywood vessels, resulting in a very high longitudinal permeability even though the vessels occupy less than 10% of the cross section. The majority of the remainder of the tissue consists of thick-walled fiber cells of extremely low permeability making it extremely difficult for liquids to penetrate except at very high pressure (Siau 1970a). The presence of parallel zones is also discussed in Chapter 7.6.1. This extreme nonhomogeneity of wood structure is one of the principal reasons why it is difficult to correlate the steady-state gas permeability of wood with its treatability with preservative liquids which involves unsteady-state transport.

3.19 General Permeability Variation with Species

Permeability is an extremely variable property of wood. Smith and Lee (1958) measured the longitudinal air permeability of approximately 100 species and found a range of values with a ratio of $5 \times 10^6:1$ for hardwoods and $5 \times 10^5:1$ for softwoods. Similar ranges of values would be expected for transverse flow, but experimental difficulties prevent the measurement of extremely low permeabilities. Table 3.2 reveals approximate values for a few common classifications of wood. The red oaks may have permeabilities as high as 200 darcys because of their large earlywood vessels. American basswood (*Tilia americana*) is very permeable because of its open, diffuse-porous structure. Pine sapwoods are among the most permeable softwoods and may have values as high as 1 to 8 darcys. The spruces and cedars are usually much lower, in the 10^{-1} darcy range. One of the lowest is Rocky Mountain (intermountain) Douglas-fir (*Pseudotsuga menziesii* var. *glauca*) heartwood in the range of 10^{-3} darcy, making it extremely refractory to impregnation.

Comstock (1970) summarized ratios of longitudinal-to-tangential permeabilities of softwoods, measured by several investigators, between 500 and 80,000 to one with most of these above 10,000 in reasonable agreement with his model. On the other hand, longitudinal-to-radial ratios extended from 15 to 50,000 to one. This greater variability in the radial direction could be related to large differences in the permeability of ray tissues. With impermeable rays a very high ratio would be expected because most of the pits are on the radial surfaces of the tracheids resulting in low radial permeability. Highly conductive rays would give rise to a low ratio.

Table 3.2. Approximate values for some common classifications of wood

Permeability, darcys	Longitudinal permeabilities	
100	10^2	Red Oak $r \approx 150\ \mu m$
10	10^1	Basswood $r \approx 20\ \mu m$
1	10^0	Maple, Pine sapwood, Douglas-fir sapwood (Pacific coast)
0.1	10^{-1}	Spruces (Sapwood)
		Cedars (sapwood)
0.01	10^{-2}	Douglas-fir heartwood (Pacific coast)
		White Oak Heartwood
		Beech heartwood
0.001	10^{-3}	Cedar heartwood
		Douglas-fir heartwood (Intermountain)
0.0001	10^{-4}	
0.00001	10^{-5}	Transverse Permeabilities (The species are in approximately the same order as those for longitudinal permeabilities).
0.000001	10^{-6}	

In hardwoods, Comstock (1975 unpublished results) found little difference between tangential and radial permeability and tabulated L/T ratios measured by several workers with a range between 30,000 and 4×10^8 to one with the highest ratio for the ring-porous red oaks. Smith and Lee (1958) found 65,000 to one in *Fagus sylvatica* and Kininmonth (1971) measured 50,000 to one in *Nothofagus fusca* and 1×10^6 in *Eucalyptus regnans*. These high ratios in hardwoods may be explained by: (a) the generally poorer penetrability of rays in hardwoods than in softwoods, and (b) the very high longitudinal permeability of ring-porous hardwoods with open vessels.

Chapter 4
Capillarity and Water Potential

4.1 Surface Tension

Surface tension is a characteristic of a liquid–gas interface. It may be attributed to an imbalance of intermolecular forces of attraction (van der Waals' forces) as illustrated in Fig. 4.1. Within the body of the liquid the intermolecular forces are essentially in balance while, on the surface, there is no upward component. Therefore, there is a net downward force, normal to the surface, and the surface is apparently in tension. If a line is visualized along the surface of a liquid, the forces on one side are balanced by those on the other and the surface tension at the interface is equal to the magnitude of the total force along one side of the line divided by the length. Surface tension is then expressed as force per unit length.

$$\gamma = F/x, \tag{4.1}$$

where γ = surface tension at interface, dyne/cm, F = force along a line of length x, dyne, x = length, cm.

Surface tension may also be defined as surface energy per unit area, or *specific surface energy*. Because of surface tension, a quantity of liquid will seek the smallest possible surface area representing the lowest energy state. Thus, a droplet or bubble is spherical in the absence of other forces. Work is required to increase the area of an interface and all the energy can be recovered when the area is reduced again. Therefore, the energy stored in a surface is potential energy similar to the energy in a stressed spring. The units for specific surface energy are those of energy per unit area, or ergs per square centimeter, which are equivalent to dynes per centimeter. Stated mathematically in derivative form,

$$\gamma = dW/dA, \tag{4.2}$$

where γ = surface tension or specific surface energy, erg/cm^2, dW = work, erg, required to increase the area by dA, cm^2.

4.2 Capillary Tension and Pressure

Assume the existence of a spherical bubble in a liquid, as illustrated in Fig. 4.2. Work must be expended to enlarge the bubble by pushing down on the piston. The quantity of work may be calculated as the product of the pressure differential and the increase in the volume. According to the definition of specific surface energy,

$$\gamma = \frac{dW}{dA} = \frac{(P_0 - P_1)dV}{dA},$$

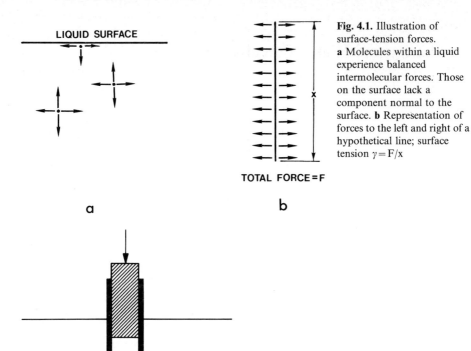

Fig. 4.1. Illustration of surface-tension forces. **a** Molecules within a liquid experience balanced intermolecular forces. Those on the surface lack a component normal to the surface. **b** Representation of forces to the left and right of a hypothetical line; surface tension $\gamma = F/x$

Fig. 4.2. Illustrating the enlargement of a gaseous bubble in a liquid by doing pressure-volume work to enlarge the area of the liquid–gas interface

where P_0 = pressure in gaseous phase, dyne/cm², P_1 = pressure in liquid phase adjacent to interface.

Since $A = 4\pi r_i^2$ and $dA = 8\pi r_i dr_i$,

$$P_0 - P_1 = \frac{2\gamma}{r_i}, \tag{4.3}$$

where r_i = radius of gas–liquid interface.

The capillary pressure equation may also be derived from the forces acting on a liquid which has risen or fallen in a capillary tube as depicted in Fig. 4.3. A liquid rises in a capillary tube when it wets the walls of the tube due to a strong force of adhesion between the liquid and the wall. An example of this is water in contact with glass or wood. In such cases the upward-acting surface-tension force must be balanced by the downward force due to the pressure differential, $P_0 - P_1$. Then

$$2\pi r \gamma \cos\theta = (P_0 - P_1) \pi r^2$$

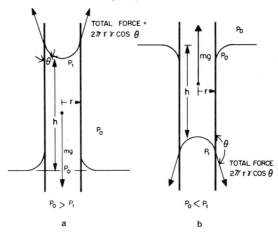

Fig. 4.3. Capillary rise and depression. **a** Liquid wets capillary; **b** liquid does not wet capillary

or

$$P_0 - P_1 = \frac{2\gamma \cos\theta}{r}, \tag{4.4}$$

where θ = wetting angle, r = radius of capillary.

Equation (4.4) is known as Jurins' Law. It is clear that, for a wetting angle of zero, Eq. (4.4) is identical to Eq. (4.3) which was derived based upon a spherical bubble. When θ is between 0° and 90°, the radius of the meniscus, r_i, is greater than the radius of the capillary, r. It is clear from Eqs. (4.3) and (4.4) that

$$r_i = r/\cos\theta. \tag{4.5}$$

In addition, the upward surface-tension force must balance the weight of the liquid in the tube. Then

$$2\pi r \cos\theta = w\, g,$$

where w = mass of liquid, g, g = acceleration of gravity = 980 cm/s². Expressing mass in terms of density, ϱ, and volume, V,

$$2\pi r \cos\theta = \varrho\, V\, g = \pi\, r^2\, \varrho\, g\, z,$$

where z = height of capillary rise, cm.
Then

$$\gamma = \frac{r\varrho g z}{2\cos\theta}. \tag{4.6}$$

Equation (4.6) may be used to calculate surface tension from the height of rise in a capillary. The contact angle is less than 90° for liquids which wet the capillary wall, and for water in contact with glass the angle is approximately zero. If a glass capillary tube is placed in mercury there is no wetting, and the contact angle is approximately 130°. Since the cosine of this angle is -0.64 and the value of z is negative, a positive surface tension is calculated by Eq. (4.6). With nonwetting liquids the meniscus is concave toward the liquid, the contact angle is between 90° and 180°, and P_1 exceeds P_0. With wetting liquids, on the other hand, the meniscus is convex toward the liquid, and P_0 exceeds P_1. It is clear from this that *the pressure is always highest on the side of a meniscus where the center of curvature is located.*

Equation (4.4) may be simplified for the convenient calculation of the capillary-pressure differential across a water–air interface in a circular capillary. Assuming a wetting angle of 0° and $\gamma = 73$ dyne/cm,

$$P_0 - P_1 = \frac{1.46 \text{ atm } \mu\text{m}}{r}, \qquad (4.7)$$

where r = radius, μm.

It is clear from Eq. (4.7) that, for a radius of 10 µm, which could be typical of a softwood tracheid, $P_0 - P_1 = 0.146$ atm while, for a radius of 1 µm, typical of a large pit opening, $P_0 - P_1 = 1.46$ atm. If P_0 is equal to 1 atm, then P_1 is -0.46 atm. This negative pressure is called *capillary tension*. With smaller radii, much higher tensions are possible which can explain such phenomena as the collapse of wood during drying, pit aspiration, and the rise of sap in trees. These topics will be discussed later. While the liquid pressure (P_1) may have negative values, the pressure on the gaseous side of the meniscus (P_0) must always be positive. Gas pressures may be reduced to very low values with a vacuum pump, but they cannot become negative and there will always be some positive absolute pressure in the system.

The pressure difference across an air–water capillary meniscus for various values of r, according to Eq. (4.7), is presented in Fig. 4.4.

When Eq. (4.4) is applied to mercury in wood, Stayton and Hart (1965) assumed a contact angle of 130°. When the surface tension is taken as 435 dyne/cm,

$$P_1 - P_0 = \frac{6.12 \text{ atm } \mu\text{m}}{r}. \qquad (4.8)$$

Accordingly the pressure difference across the mercury–air meniscus is revealed in Fig. 4.5.

4.3 Mercury Porosimetry

Practical use has been made of Eq. (4.8) for the determination of the pore-size distribution of wood and other porous solids. Stayton and Hart (1965) forced mercury into short specimens of *Pinus taeda*, *Picea engelmannii*, and *Chamaecyparis lawsoniana* using a porosimeter with a capacity of 3,000 lb/in² (204 atm). Figure 4.5 indicates that this pressure will fill capillaries with radii as small as 0.03 µm. Their procedure was to initially evacuate the specimen and then gradually increase the pressure while measuring the volume of mercury injected into the wood. These pressure-volume data could then be utilized to calculate the pore size distribution on a volume-fraction basis. In this method specimens are usually cut to less than cell length to permit filling of all the lumens. Similar determinations were made by Schneider and Wagner (1974) using several species of wood, densified wood (lignostone), and wood saturated with polyethylene glycol. In this study it was assumed that $\theta = 141.3°$ and $\gamma = 480$ dyne/cm resulting in slightly higher values of r than those calculated from Eq. (4.8).

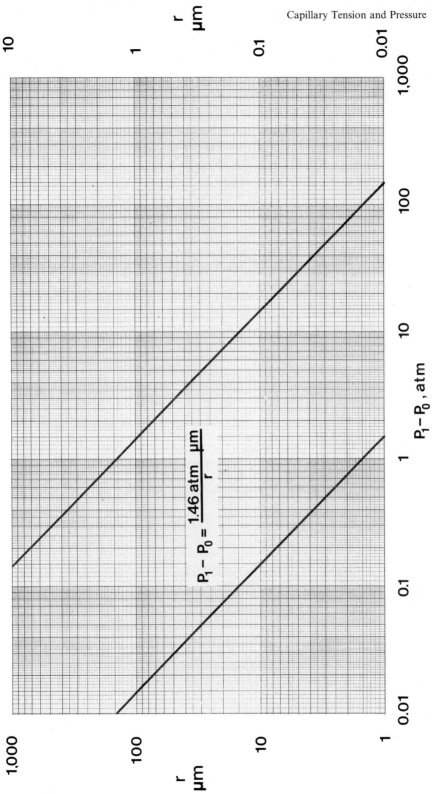

Fig. 4.4. Pressure differential $(P_0 - P_1)$ or negative water potential $(-\psi)$ in atmospheres across a water–air meniscus of radius, r, in micrometers, assuming $\gamma = 73$ dyne/cm and $\theta = 0°$. $P_0 - P_1 = 1.46$ atm μm/r

110 Capillarity and Water Potential

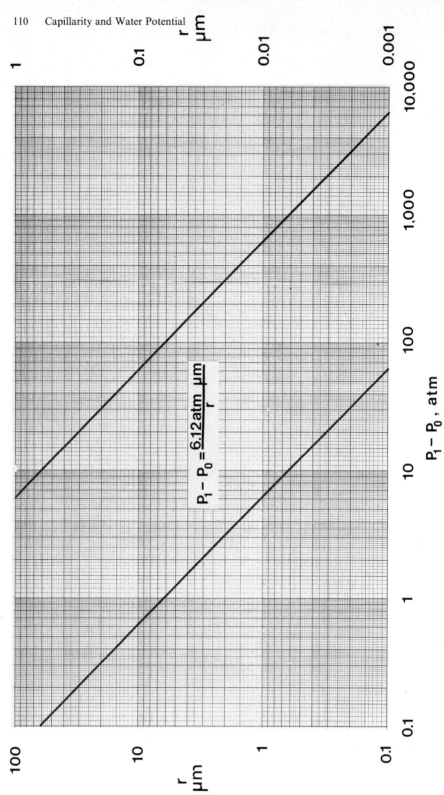

Fig. 4.5. Pressure in atmospheres $(P_1 - P_0)$ required to force mercury into a circular capillary of radius, r, in micrometers, assuming $\gamma = 435$ dyne/cm, $\theta = 130°$. $P_1 - P_0 = 6.12$ atm μm/r.

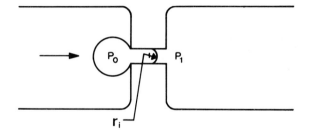

Fig. 4.6. Illustrating the passage of an air bubble through a pit opening

4.4 Influence of Capillary Forces on the Pressure Impregnation of Woods with Liquids

Comstock (1965) and Kelso et al. (1963) describe the precautions necessary in the measurement of the liquid permeability of wood to prevent decreased flow due to air blockage or the presence of particulates in the liquid. Comstock was able to show that the specific permeability measured with a gas (extrapolated to zero reciprocal pressure) was equal to that measured with a nonswelling liquid. Hudson and Shelton (1969) found that the flow rate of water-borne preservatives into green southern pine poles by the Pres-Cap process could be increased up to 400-fold by cutting disks from the butt end after an initial application of pressure. This increase was attributed to the elimination of air between the preservative solution and the sap column in the log. An explanation of these phenomena may be clarified by reference to Fig. 4.6. Air bubbles are always present either within a liquid or within the lumens of the wood cells, and these must ultimately be forced through minute pit openings. When this occurs, the spherical bubble must be deformed at the leading surface, resulting in an increase in surface area and the expenditure of energy in the form of increased pressure. Although the contact angle within the pit opening is probably near zero, the leading surface will become hemispherical with a radius approximately equal to that of the opening. The pressure, P_0, inside the bubble must then exceed P_1 in the liquid in front of the meniscus by an amount which may be calculated from Eq. (4.4). If the preservative is a water-borne solution, an approximate pressure can be determined from the simplified Eq. (4.7) or Fig. 4.4 with the assumption that the contact angle is zero and that the surface tension has the same value as water.

4.5 Collapse in Wood

Kauman (1958) investigated the causes of collapse during the drying of *Eucalyptus regnans*. He found that drying stresses were a significant factor, since the shell of a board undergoing drying shrinks more than the core initially, exerting a compressive stress upon it which can contribute to collapse. Capillary tension was recognized, however, as the principal cause of collapse.

Table 4.1. Surface tension of liquid-air interfaces at 20 °C

Liquid	Surface tension, dyne/cm
Acetone	23.32
Benzene	28.88
Carbon tetrachloride	28.6
Ethanol	22.32
Mercury	485.
Methanol	22.6
Pentane	16.
Toluene	28.55
Water	72.75

It is clear that the capillary tension resulting from the drying of wood can be reduced by the displacement of the water in wood by a liquid with lower surface tension. Ellwood et al. (1959) have demonstrated that collapse can be eliminated in the drying of susceptible hardwoods by exchanging the water with methanol and ethanol. The surface tensions of several liquids are given in Table 4.1, where it is seen that many organic solvents have values of approximately one-third that of water. While collapse can frequently be prevented by such solvent-exchange procedures, it is not economically feasible on a commercial scale.

Collapse occurs in wood when the capillary tension exceeds the compressive strength perpendicular to the grain. In the case of a susceptible species such as *Sequoia sempervirens*, the Wood Handbook (USDA 1955) gives a value of 520 lb/in^2 (35.4 atm) for green material. Reference to Fig. 4.4 indicates that collapse may occur if the radii of the pit openings are smaller than 0.04 µm.

A description of the mechanism of collapse in wood due to drying is provided here through the courtesy of C. Skaar (1970, personal communication). The limiting factor in the elimination of lumber degrade during drying is often associated with the removal of free or capillary water in the cavities of the wood. Low temperatures must often be used during the early stages of drying because of the hazards associated with too-rapid removal of capillary water at higher temperatures. The key to future progress in speeding up the drying process appreciably is probably to be found in discovering new methods of removing capillary water more effectively than is possible with present drying methods. Therefore, it is important to understand the movement of capillary water during the drying process as it relates to the onset of capillary tension within the wood. The principles described below were elucidated by Hawley (1931).

The forces causing water to rise in a capillary tube are revealed in Figs. 4.3a and 4.7a. The surface tension at the water–air interface causes a tension pull (shown by the upward-pointing arrow) on the water immediately below it. At equilibrium, the water exerts an equal tension on the surface in the opposite direction, as indicated by the heavy arrow. This water tension pulls inward on the walls of the tube as well as on the surface. The effect of gravity will be neglected because in most drying situations it is negligible compared with capillary forces.

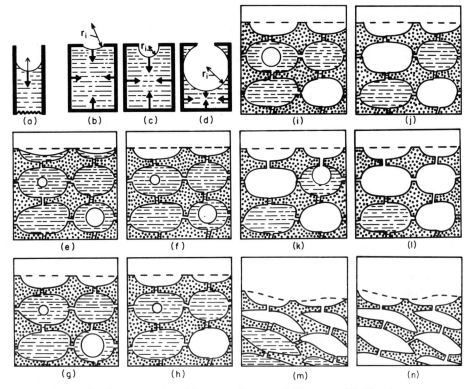

Fig. 4.7. Illustrating the evaporation of free water from wood (Courtesy of C. Skaar)

Figure 4.7b shows a capillary surface from which liquid water is evaporating into the air. The radius of curvature, r_i, is large at this stage, and therefore the capillary tension is small. As evaporation continues, the radius of curvature decreases until it reaches the radius of the evaporating hole (Fig. 4.7c). The capillary tension increases correspondingly, as indicated by the increased length of the heavy arrows. Continued evaporation results in an *increase* in the capillary radius and therefore a *decrease* in the capillary tension (Fig. 4.7d). It is clear that maximum tension occurs when the evaporating radius becomes equal to that of the opening, as seen in Fig. 4.7c. Figures 4.7e to 4.7l show how removal of capillary water is thought to occur in green wood. The lumens are initially completely water-filled except for two with air bubbles of different sizes (Fig. 4.7e).

Drying takes place only from the upper surface of the wood which is exposed to the air. Wood cells enclose the other three sides of the diagram. As capillary water evaporates from the exposed cells, curved menisci form on the air–water interfaces. The capillary tension is small at this stage of drying, due to the relatively large radii of the drying surfaces.

When the surface cells are empty, the evaporating surfaces appear in the pit openings (Fig. 4.7f). The evaporating surface radii are reduced and approach the radius of the large bubble. As the radii become smaller, the capillary tension in the cell system increases. This tension acts on the cell walls and on the air–water menisci throughout the system, causing expansion of the air bubbles. Since the tension around the larger bubble is lower, it expands first and the cell is emptied (Fig. 4.7h). The water from the cell migrates principally through the adjoining cell to the evaporating surface. Therefore, it is possible for a cell deep in the wood to lose its capillary water earlier than one closer to the surface. This may explain the presence of wet pockets in wood after drying.

After the large bubble has expanded, filling the cell, the surface curvatures are again reduced as the menisci retreat into the pit openings. Once more, the tension in the system increases until it is great enough to expand the smaller bubble (Fig. 4.7i) causing the displaced water to migrate through the adjoining cell wall and pit opening and finally to evaporate from the surface (Fig. 4.7j).

As evaporation continues and the meniscus retreats into the pit openings, tension in the capillary water mounts even higher than before, since there are no air bubbles nearby to relieve the tension by expanding. The maximum tension exerted on the cell walls is determined by the size of the largest opening in the system, in accordance with Eq. (4.4). Most woods are able to withstand this maximum capillary tension. When the evaporating surface reaches the full cell, it expands into the cell (Fig. 4.7k). This results in a gradual increase in radius of curvature, accompanied by a reduction in capillary tension. The water evaporates into the air space, and migrates out to the wood surface in vapor form, thus emptying the cell (Fig. 4.7 l). Other cells deeper in the wood lose their capillary water in a similar manner.

Capillary tension can cause collapse in wood while the meniscus moves through the pit openings if these are sufficiently small and if the cell wall is thin, resulting in low compressive strength perpendicular to the grain. A single cell or a group of cells may collapse (Figs. 4.7m and n). This effect is more pronounced at high temperatures, because the cell wall is significantly weakened as temperature increases. Collapsed wood can often be restored to the original shape by subjecting it to high temperature and humidity for a period of time after it has been dried. In this case the air and water vapor in the wood exert a pressure inside the cell which reverses the tension forces that originally caused the collapse.

The factors responsible for collapse of wood during drying are listed below:

1. Small pit-membrane openings increase the susceptibility of the wood to collapse due to the resulting high capillary tension. Softwoods with small pit openings are characterized by low permeability.
2. A high surface tension of the liquid evaporated from the wood promotes collapse. When free water is replaced by a low-surface-tension organic liquid, collapse may sometimes be prevented.
3. A low-density wood has thin cell walls which collapse easily due to low compressive strength.
4. Elevated temperatures decrease strength and therefore render the wood more susceptible to collapse.

Examples of woods susceptible to collapse are heartwood of *Sequoia sempervirens*, *Thuja plicata*, *Tsuga heterophylla*, *Juglans nigra* and many species of *Eucalyptus*.

4.6 Pit Aspiration

Aspirated pits reduce the permeability of wood, thus increasing the difficulty of fluid impregnation. Several investigations have related the extent of pit aspiration with the capillary tension resulting from the removal of free water or organic solvents from wood. Phillips (1933) conducted a microscopic study of *Pseudotsuga menziesii* and *Pinus sylvestris* to observe the condition of the pit membranes during drying. He found most of the pits unaspirated in green sapwood, with the proportion of unaspirated pits gradually decreasing with decreased moisture content until the fiber saturation point was reached, whereupon most of the earlywood pits were aspirated. There was no significant further aspiration below the fiber saturation point. The fraction of unaspirated pits in dry latewood was significantly higher than in earlywood, thus accounting for the generally higher permeability and easier treatability of the latewood of softwoods. The fraction of unaspirated pits was approximately proportional to the specific gravity and, consequently, to the double-cell-wall thickness. The greater resistance to aspiration of latewood pits was attributed to the thicker cell walls and consequently more rigid pit membranes and to the smaller diameter of latewood pits, resulting in a much more rigid structure. A large fraction of aspirated pits was found in heartwood before drying when the moisture content was below the fiber saturation point. Phillips soaked green wood in alcohol and found no additional aspiration when the alcohol was dried from the wood. This suggested that the lower surface tension of alcohol was sufficient to prevent aspiration because of the reduced capillary tension.

Bolton and Petty (1977a, b) investigated the effect of solvent-exchange drying and critical-point drying on the permeability of wood specimens of outer sapwood of *Pinus sylvestris*. They found that the permeability of the critical-point-dried material was the highest of all, and attributed this to unaspirated pits as a result of the absence of a liquid–air interface during drying. The sample dried from water had the lowest permeability and those with organic solvents with surface tensions of 16 and 34 dyne/cm had intermediate permeabilities. Therefore the permeability decreased as the surface tension of the liquid from which the wood was dried increased, the decrease being explained by an increase in pit aspiration as the surface tension forces were increased.

Thomas and Nicholas (1966) dried *Pinus taeda* specimens from water, methanol, and pentane, and were able to eliminate pit aspiration with the methanol and pentane although it occurred in specimens dried with water. The pentane-dried specimens were initially solvent-exchanged through methanol and acetone. It was possible to reverse the aspiration process by resaturating the specimens, exchanging the water with organic solvents, and redrying. Electron micrographs then indicated that the pits were deaspirated.

Liese and Bauch (1967b) studied the influence of drying on the aspiration of pits of *Picea abies*, *Pinus sylvestris*, *Abies alba* and *Thuja occidentalis*. Green wood was exchanged with water, ethanol, acetone, and mixtures of water and these solvents. Water-penetrability tests were followed by microscopic examination. The pits of *Thuja occidentalis* have no definite tori and are therefore not subject to aspiration. The water permeability of this species was essentially unaffected by drying from water. The pits in the earlywood of the other species were all aspirated when dried from water and were not aspirated when dried from alcohol or acetone. A variable proportion of the pits of latewood were not aspirated when dried from water, depending upon the species. Some aspiration occurred when specimens were dried from 75% ethanol-water and 80% acetone-water, both mixtures having a surface tension of 26 dyne/cm, and it was concluded that the surface tension must be at least this high to produce aspiration of the pits of earlywood.

Comstock and Côté (1968) compared air drying with solvent-exchange drying of *Pinus resinosa* and *Tsuga canadensis*, using several liquids including surfactant solutions in water with surface tensions of approximately 20 dyne/cm (compared with 73 dyne/cm for water) and with a series of organic solvents with surface tensions between 17 and 44 dyne/cm. Except for ethanol, methanol, and acetone, the solvents were nonpolar and immiscible with water. Therefore an intermediate exchange was required using either ethanol or acetone. Permeability was measured before drying with water, with 5.3 darcys obtained for both species. The specimens dried from organic solvents were essentially unchanged from these values, even when dried from furfural with a surface tension of 44 dyne/cm. The air-dried *Pinus resinosa* wood had a permeability of approximately 10% of the undried value while that of *Tsuga canadensis* was 0.3%. Although the surfactant-water solutions had low surface tensions, solvent-exchange drying from these solutions resulted in greatly reduced permeabilities due to pit aspiration. It was concluded from this that only solvents which do not promote adhesion between the torus and border, presumably by hydrogen-bonding forces, are effective in preventing pit aspiration.

Thomas and Kringstad (1971) studied several factors influencing pit aspiration in never-dried earlywood of *Pinus taeda* sapwood. They found that chemical blocking of hydroxyl groups by esterification and etherification prevented complete aspiration of 60% of the pits after drying from water, while all of the pits were completely aspirated when no chemical treatment was used. This indicated the important role of hydroxyl groups in pit aspiration. The effect of drying from several liquids on the extent of pit aspiration was studied. Aspiration occurred on drying from water-surfactant solutions of low surface tension but not with some organic liquids of high surface tension, in agreement with the findings of Comstock and Côté. The results of this investigation indicated that the evaporating liquid must possess all of the following properties in order to cause pit aspiration: (a) the evaporating liquid must be capable of forming hydrogen bonds and must have both donor and acceptor properties, (b) the liquid must swell wood nearly as much as or more than water, and (c) the surface tension and contact angle relationship must be such that the initial capillary force is sufficient to cause pit-membrane displacement. It was found additionally that deaspiration could be accomplished by soaking in water to break the hydrogen bonds between the pit membrane and border, followed by drying from pentane after solvent exchange. As the length of time in the aspirated state increased, a longer soaking time was required to produce deaspiration.

A mechanism for pit aspiration due to capillary tension was proposed by Hart and Thomas (1967) (as illustrated in Fig. 4.8 a to d). In Fig. 4.8 a, a meniscus of large radius was formed in the pit aperture. In (b) an annular meniscus has formed between the pit border and torus with greater tension than in (a). After the upper pit cavity has emptied, evaporation takes place from the small openings in the margo (c), which produce a much higher tension capable of forcing the membrane

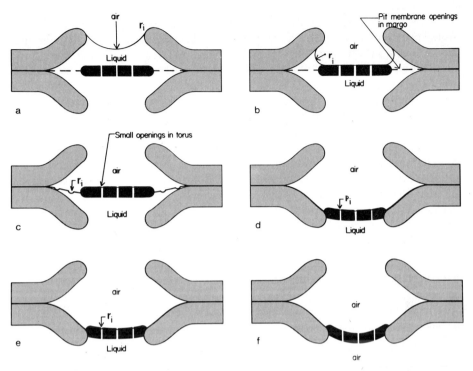

Fig. 4.8a–f. Proposed mechanism of pit aspiration (**a** through **d**), according to Hart and Thomas (1967) with an additional proposed mechanism (**e** and **f**) to explain dishing of the torus. (According to Nicholas 1970 personal communication). **a** Meniscus in pit aperture; **b** annular meniscus between pit border and torus; **c** menisci in pit-membrane openings; **d** aspirated pit with relatively large value of r_i in torus openings; **e** torus becoming dished due to capillary force resulting from a small r_i in torus openings; **f** completely aspirated pit with dished torus

toward the lumen filled with water. This process continues until the pit is aspirated, with the flat torus covering the aperture as in (d). There is evidence of the dishing of the torus inward toward the aperture (see Figs. 2.14, 2.15, and 2.19). This could be caused by capillary forces due to very small openings in the torus, according to Nicholas (1970 personal communication). Such openings are generally invisible, but could exist as tortuous paths between the randomly oriented microfibrils of the torus similar to the openings between the fibers of filter paper. Since the openings are smaller than those in the margo, the resulting capillary force is greater. The dished torus is illustrated in Fig. 4.8e, and the fully aspirated pit with maximum dishing in 4.8f.

Bolton and Petty (1978) proposed a model for the determination of the force required to deflect the pit membranes of the wood of *Pinus sylvestris*. Their results indicated that a pressure of 5.4 atm is required in first-formed earlywood and 680 atm in the center of the latewood. This difference is indicative of the relative rigidities of the two kinds of membranes and explains the greater resistance to aspiration of latewood pits along with the generally higher treatability of the latewood of softwoods. If water with a surface tension of 72 dyne/cm were dried from the wood, pit openings with maximum radii smaller than approximately 0.3 μm would be required to deflect the earlywood membrane and those to deflect the latewood membranes must be smaller than 0.002 μm. The former value is relatively large for a pit opening and therefore aspiration in the earlywood may be expected while the latter value is very small resulting in a greatly reduced probability of aspiration in the latewood. Bolton and Petty pointed out in 1980 that the surface tension of sap could be as low as 40 dyne/cm corresponding to smaller radii required cause aspiration during drying.

In a closely related field of research, Stamm (1964) described the measurement of maximum-effective pit radii by forcing air through saturated wood and calculating the radii from measured bubble points. Bolton and Petty (1980) have pointed out that an error could result in this technique in the event of total membrane displacement, causing the torus to seal off the aperture at a pressure lower than that required to force bubbles through openings in the membrane.

The factors which contribute to pit aspiration in wood are summarized below:

1. The evaporating liquid must be capable of forming hydrogen bonds and must have both donor and acceptor properties.
2. The evaporating liquid must swell wood nearly as much as or more than water.
3. The pit-membrane openings must be small.
4. The surface tension and contact angle relationship must be such that the initial capillary force is sufficient to cause pit-membrane displacement.
5. The pit membrane must have low rigidity, which is characteristic of low-density earlywood of softwoods.

4.7 The Relationship Between Water Potential and Moisture Movement

In Chap. 6, Fick's first law of diffusion will be discussed in which the steady-state movement of moisture in wood below the fiber saturation point is attributed to a gradient of moisture concentration or wood-moisture content. This is suitable under very specific and isothermal conditions. In a more general way, water potential (ψ) may be regarded as the driving force for the transport of water in both liquid and vapor phases including bound water in wood. The water potential of a level surface of pure water at 0 °C may be taken as zero. If this surface is located in a confined space, permitting the air to be saturated with water vapor, a state of equilibrium will be achieved when the water potentials of both the liquid and gaseous phases are equal. Therefore, the water potential of saturated vapor at 0 °C is also zero.

The water potential in a system has several components: (a) ψ_s due to reduced vapor pressure over a solution including wood below the fiber saturation point, and of the unsaturated air in equilibrium with either of these, (b) ψ_p due to a static pressure difference, (c) ψ_g due to a difference in gravitational level, and (d) ψ_m due to matric potential caused by differences in capillary pressure.

The component ψ_s, characteristic of unsaturated air or wood below the fiber saturation point, is usually the most significant in wood-moisture relationships. It may be defined as:

$$\psi_s = \psi_1^0 + \frac{RT}{18 \text{ cm}^3/\text{mol}} \ln h, \qquad (4.9)$$

where ψ_1^0 = water potential of saturated vapor in equilibrium with a level surface of pure water at 1 atmosphere and at temperature T, with ψ_1^0 taken as zero at 0 °C, R = universal gas constant = 82 (atm cm³)/(mol K), h = relative vapor pressure = p/p_0.

The water potential of moist air at 0 °C as a function of the relative vapor pressure is illustrated in Fig. 4.9. It is evident from this that very high negative values result even from moderately high values of h indicating the presence of high gradients which are available to cause the evaporation of moisture from a level water surface, from saturated wood, or from the surface of a leaf during transpiration. In the case of the drying of wood, equilibrium will occur when the water potential of the wood becomes equal to that of the surrounding air. This occurs at the equilibrium moisture content of the wood.

The term ψ_1^0 in Eq. (4.9) is taken as zero at 0 °C and it becomes increasingly positive as the temperature rises. Its value does not affect the gradient in an isothermal system because it has the same value throughout. However, in a nonisothermal situation its contribution to the overall gradient may be substantial as will be discussed in Chap. 6.8. The values of ψ_1^0 may be calculated from the corresponding temperature-dependent term of chemical potential, μ_1^0, as obtained from Heimburg (1981, personal communication) and from data provided by Castellan (1966) in units of cal/mol. Chemical potential may be easily converted to water potential by dividing the former, expressed in energy per mole, by the molar volume of water, 18 cm³ per mole, and by expressing R in units of (atm cm³)/mol K rather than cal/(mol K), yielding water potential in units of pressure. The term ψ_1^0 may be calculated as

$$\psi_1^0 = 23.77 \text{ C} + 0.018 \text{ C}^2, \qquad (4.10)$$

where ψ_1^0 is expressed in atm, C = Celsius temperature.

When Eqs. (4.9) and (4.10) are expressed in terms of chemical potential (μ), they assume the forms:

$$\mu = \mu_1^0 + RT \ln h, \qquad (4.11)$$

where μ = chemical potential, cal/mol, R = universal gas constant = 1.987 cal/(mol K), and

$$\mu_1^0 = 10.37 \text{ C} + 0.0077 \text{ C}^2, \qquad (4.12)$$

where μ_1^0 = chemical potential of saturated vapor in equilibrium with a level surface of pure water at temperature C, cal/mol.

The component due to gravitational level, ψ_g, is significant where there is a difference in head of a water column. It may have an important effect on the rise of sap in tree stems.

The gravitational component may be calculated as

$$\psi_g = \varrho \, g \, z / [1.013 \times 10^6 \text{ dyne}/(\text{cm}^2 \text{ atm})], \qquad (4.13)$$

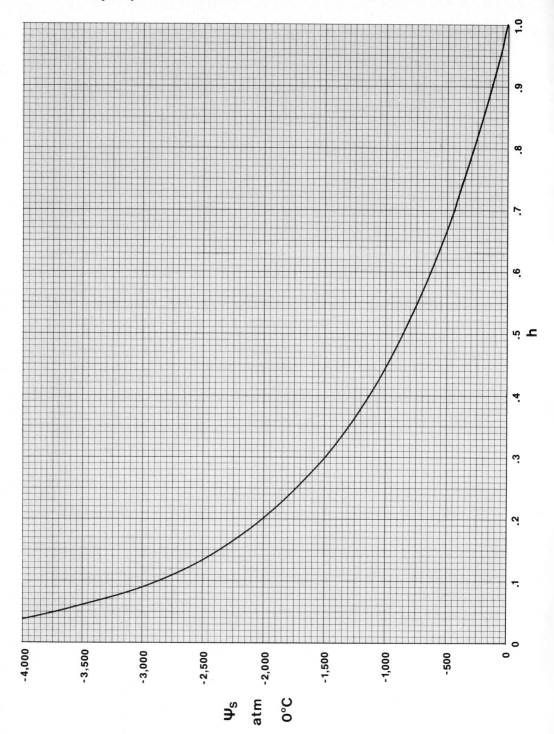

Fig. 4.9. Water potential (ψ) in atmospheres of unsaturated air vs. relative vapor pressure (h) at 0 °C

where ϱ = density of water, g/cm^3, g = gravitational acceleration = 980 cm/s^2, z = difference in gravitational level, cm.

The matric component of water potential is equal to the difference between the pressures in the liquid and gaseous phases on both sides of a liquid–gas meniscus as defined by Eq. (4.4). Its value will always be negative in a wetting liquid where the contact angle is less than 90°.

$$\Psi_m = -\frac{2\gamma\cos\theta}{r \times 1.013 \times 10^6 \text{ dyne}/(\text{cm}^2 \text{ atm})}. \tag{4.14}$$

It is evident from Eq. (4.14) that a negative water potential exists in the free water within the capillary structure of wood. Therefore vapor can migrate from saturated air into the capillary structure of wood where it will condense as free water. This may occur only at very high relative humidities where the water potential of the moist air is more positive than that in the capillaries. Therefore, in theory, wood will become totally saturated with free water when it is exposed to saturated air with a relative humidity of 100%.

The relationship represented by Eq. (4.14) may be simplified by assuming a value of γ of 73 dyne/cm and a contact angle of 0° for free water in wood. Then,

$$\psi_m = -\frac{1.46 \text{ atm } \mu m}{r}, \tag{4.15}$$

where r = radius of capillary, μm.

This is identical with Eq. (4.7) except for the substitution of ψ_m for $P_0 - P_1$ and the negative sign. Therefore, a plot of $-\psi_m$ vs r may be found in Fig. 4.4.

As a consequence of the reduced water potential in a filled capillary, the water potential of the moist air in equilibrium with its surface is also reduced, resulting in a lower vapor pressure at saturation. The relative vapor pressure over a capillary meniscus may be calculated by combining Eqs. (4.9) and (4.14). Assuming isothermal conditions and $\theta = 0°$,

$$\ln h = \frac{36\gamma}{rRT}, \tag{4.16}$$

where $R = 8.31 \times 10^7$ erg/(mol K).

Equation (4.16) may be simplified by substituting values of 73 dyne/cm for γ, and 313 K (40 °C) for T. Then

$$\ln h = -1.01 \times 10^{-3} \, \mu m/r, \tag{4.17}$$

where r is expressed in μm.

The term ln h approaches (h-1) at high values of h. If h > 0.95, it may be stated that ln h = (h – 1). Then, solving Eq. (4.17) for r,

$$r = -1.01 \times 10^{-3} \, \mu m/\ln h \tag{4.18a}$$

or,

$$r = 1.01 \times 10^{-3} \, \mu m(1-h), \tag{4.18b}$$

h > 0.95, T = 40 °C.

Table 4.2. Values of the matric component of water potential for various radii with corresponding values of equilibrium relative humidity at 40 °C

r, μm	Equilibrium relative humidity, H %	ψ_m, atm
0.01	90.5	− 144
0.05	98.0	− 27.6
0.1	99.0	− 14.4
1	99.9	− 1.44
1.5	99.93	− 1.0
4	99.97	− 0.36
10	99.99	− 0.14
150	99.9999	− 0.010

From Eq. (4.16) it follows that the component ψ_s for wood in equilibrium with unsaturated air, may be considered as a matric component corresponding with radius, r. Thus it is difficult to distinguish between the components ψ_s and ψ_m in wood which complicates the problem of finding a clear division between bound and free water. Therefore it is proposed to represent the water potential of wood by the symbol ψ. When the temperature-dependent term in Eq. (4.9) is neglected in the isothermal case, the relative vapor pressure may be calculated from the water potential using the following relationship,

$$\ln h = \frac{18 \text{ cm}^3/\text{mol}(\psi)}{RT} \tag{4.19}$$

where $R = 82$ (atm cm³)/(mol K).

Equation (4.19) may be simplified by making substitutions similar to those applied to Eq. (4.16). The water potential of wood may then be calculated as a function of the relative vapor pressure as

$$\psi = 1{,}426 \ln h \text{ atm} \tag{4.20a}$$

or

$$\psi = 1{,}426 \, (h-1) \text{ atm,} \tag{4.20b}$$

$h > 0.95$, T = 40 °C.

Table 4.2 reveals the relationship between r, H, and ψ as calculated from Eqs. (4.18) and (4.20). The relationship between the reduced vapor pressure (1–h) and capillary radius is plotted in Fig. 4.10. It is clear from this that a relative humidity of 98% corresponds to a capillary radius of 0.05 μm which is in the range of the size of the openings in the pit membrane. All larger capillaries including tracheid or vessel lumens and pit chambers would be expected to be empty at this relative humidity. Even though the pit openings may be filled with water they constitute a very small fraction of the total void volume, and therefore this contribution to the total moisture content will be negligible. Consequently at this level of relative humidity, very little free water should be present and this may therefore represent a reasonable value for the FSP, although it must be recognized that no definite distinction between bound and free water is evident from Table 4.2.

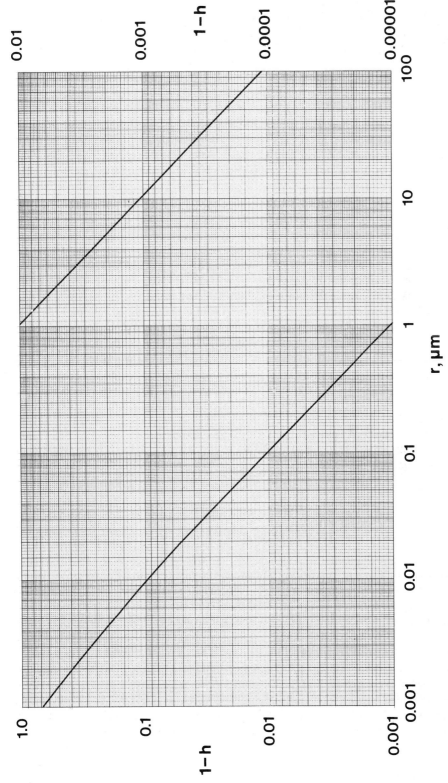

Fig. 4.10. The relationship between the radius (r) of a water-filled capillary and the reduced vapor pressure (1−h) in equilibrium with it at 40 °C

4.8 Notes on the Water Potential, Equilibrium Moisture Content, and Fiber Saturation Point of Wood

Tiemann's definition of FSP is based upon the moisture content of wood at which the cell walls are saturated with bound water with no free water in the lumens. In the light of the above discussion of water potential there appears to be no firm line of division between bound and free water, increasing the uncertainty in defining FSP. Therefore in Chap. 1.3 it was proposed that FSP be redefined as the moisture content corresponding to abrupt changes in physical properties. Stamm (1964) tabulated values of FSP of *Picea sitchensis* determined from such changes in six physical properties: (1) the sorption isotherm was obtained up to a relative humidity of approximately 98% and then extrapolated to 100% resulting in an FSP of 31%; (2) the shrinkage was extrapolated to zero shrinkage for volumetric, tangential, and radial shrinkages, giving results between 24% and 29%; (3) the apparent compression of adsorbed water obtained from the measurement of the density of the moist cell wall by immersion in benzene (as described in Chap. 1), and extrapolated to zero compression resulted in an FSP of 30%; (4) heat of wetting measurements at different moisture contents were extrapolated at zero heat of wetting, giving a value of 29%; (5) the point of deviation from linearity of the log resistivity vs. moisture content relationship, which is found at low moisture contents, produced a value of FSP at 29%; and (6) the intersection of the log mechanical strength vs. moisture content relationship with the line representing the strength of fully swollen wood was taken as the FSP, resulting in values from 26% to 30% for the various strength properties. In summary, all these methods of determination of FSP yielded values between 24% and 31%, with most of them between 27% and 31%. Therefore there is a reasonable agreement as to the moisture content corresponding to abrupt changes in all these physical properties.

Stamm (1964) recognized that FSP cannot be defined as the equilibrium moisture content corresponding to 100% relative humidity because such an equilibrium would result in total saturation of the wood. Therefore, it must be recognized that values given on a sorption isotherm plot above 98% relative humidity such as in Fig. 1.6 are incorrect and only represent the extrapolation of values measured to 98%. It is a practical reality that relative humidities above 98% cannot be controlled easily due to the very low wet-bulb depressions. It is clear from Table 4.2 that $h=0.98$ corresponds to $r=0.05$ μm and a water potential of -28 atm. Since all the lumens and pit chambers are larger than this, and since these constitute most of the void volume of the wood, this value of relative humidity would not be expected to add any significant amount of free water to the wood. In addition to this the radius of 0.05 μm is of the order of the size of the pit-membrane openings and these constitute a very small fraction of the void volume due to their small thickness of approximately 0.1 μm. Because of the relatively good agreement between Stamm's values obtained from the various physical properties and their approximate agreement with the assumed value of 30% for woods growing in the temperate zones, it is concluded that this assumption is a reasonable one, which has practical value for the wood scientist concerned with the effect of moisture content upon wood properties. It is true, however, that significantly higher values of FSP have been measured by alternative methods and these will now be discussed briefly.

Stone and Scallan (1967) determined an FSP of 40% for *Picea mariana* both by the *pressure-plate* and *nonsolvent-water* methods. The pressure-plate method, described by Robertson (1965), consists in placing a thin specimen on a porous plate with the application of positive gas pressure to the specimen which forces free water out of the wood from all capillary radii larger than that corresponding to the pressure difference indicated by Eq. (4.7). When equilibrium has been established, the moisture content is measured giving a point on the sorption isotherm in correspondence to a relative humidity which may be calculated from Eq. (4.19), (4.20), estimated from Table 4.2, or read from Figs. 4.4 and 4.10. Essentially this method consists of a direct determination from the water potential which is the negative of the applied gas pressure. The FSP was obtained by Stone and Scallan from the sorption isotherm constructed as a plot of M vs. (1–h) which is directly proportional to the water potential at $h > 0.95$, according to Eq. (4.20b). The point of inflection of this curve was selected as the FSP. They used a value of $h = 0.9975$ which corresponds to a water potential of -3.6 atm and a capillary radius of 0.4 µm. The value of 40% obtained by this method is significantly higher than that which is consistent with the volumetric shrinkage of 11.3% for bulk wood listed in the Wood Handbook (USDA 1974). The same source reveals G_f as 0.38 for this wood. Equation (1.39) therefore predicts a maximum shrinkage of 30 (0.38) or 11.4%, in close agreement with the listed value. Since Eq. (1.39) is based upon the assumption of $M_f = 30\%$, a significantly higher shrinkage would be expected with $M_f = 40\%$. Stone and Scallan measured EMC's by the pressure plate method up to $h = 0.9998$, corresponding to a water potential of -0.3 atm and $r = 5$ µm.

In the nonsolvent-water method, a dilute solution of dextran-110 of known concentration was added to saturated microtome sections of wood or wet pulp fibers. The dextran molecules are sufficiently large that they cannot penetrate the cell wall. The free water in the wood or pulp will then dilute the dextran solution. By accurate measurement of the final concentration after equilibrium has been established, the quantity of free water may be determined and the difference between this and the initial total moisture content of the saturated material is taken as the FSP. This method yielded a FSP of 42%, slightly higher than that by the pressure-plate method, but in close agreement despite the dissimilarity of the methods.

Feist and Tarkow (1967) describe the measurement of FSP by the *polymer-exclusion* method. A polymer with molecules sufficiently large that they cannot penetrate the cell wall is selected such as PEG-9000. A microtome section of saturated wood is then placed in a polymer solution of known concentration. Sufficient time is allowed for equilibrium to occur between the free water in the lumens and the external solution. After equilibrium, the wood is removed from the solution and weighed to determine the total water content. It is then extracted with water to remove all the polymer, the quantity of which is determined by analysis. The extracted wood is dried and weighed. By assuming that the concentration of polymer in the free water is the same as in the original solution, the total free water may be calculated. The FSP is then calculated from the difference between the free water and total water. The values obtained were 31% for *Picea sitchensis*, 34% for *Acer* sp., and 52% for *Achroma logapus* (balsa). These values are generally higher than the assumed 30% and the very high value for balsa ($\varrho = 0.25 \text{ g/cm}^3$) may be attributed partly to its low specific gravity.

Ahlgren et al. (1972) have also employed the polymer-exclusion technique and obtained agreement with *microscopic measurements* again made using microtome sections. In the latter case the volume fraction of swollen cell wall was determined microscopically, permitting the calculation of FSP from additional data consisting of the volume of the saturated wood, the oven-dry weight, and the specific gravity of cell-wall substance. This method gave an FSP of 53% for the wood of *Pseudotsuga menziesii*, while the polymer-exclusion method yielded 54%. Both of these values are, again, significantly higher than that expected from shrinkage data. The Wood Handbook (USDA 1974) lists a volumetric shrinkage of 12.4% and G_f of 0.45, with a predicted shrinkage of 30 (0.45)=13.5% from Eq. (1.39). An FSP of 54% would result in an expected volumetric shrinkage of 54 (0.45)=24.3%, clearly much greater than measured values for this species.

Griffin (1977) measured EMC's by the pressure-plate method over approximately the same range of relative vapor pressures as Stone and Scallan. He detected a discontinuity in the sorption isotherm at a water potential of approximately -1 atm and therefore selected this as the value for determination of the FSP. The corresponding relative vapor pressure is 0.9993 and the capillary radius 1.5 μm.

The relationship between wood-moisture content and relative humidity in equilibrium with it has been traditionally represented on linear graph paper as a sorption isotherm. This presentation is relatively satisfactory up to h=0.98. Beyond this value, the data of Stone and Scallan and Griffin indicate a rapid exponential rise in the equilibrium moisture content until full saturation with free water occurs at 100% relative humidity. An improved representation of the relationship at high relative humidities can be obtained from the moisture characteristic of wood as presented by Stone and Scallan (1967), Griffin (1977), and Baines (1980 unpublished results). Such a plot of moisture content vs. water potential as presented in Fig. 4.11. This includes an idealized plot in which ψ_s is calculated from the sorption isotherm data in the Wood Handbook (USDA 1955), down to -30 atm, corresponding to h=0.98. At higher relative vapor pressures the moisture content increases rapidly, approaching saturation at r=15 μm, $\psi=-0.1$ atm for softwoods having tracheid lumens of approximately this size. In the case of a hardwood such as red oak, saturation would be expected at approximately -0.01 atm corresponding to r=150 μm. It will be recalled that the determination of the sorption isotherm from values of EMC obtained after achieving equilibrium in a controlled humidity chamber is limited to a maximum relative vapor pressure of 0.98, corresponding to -30 atm and r=0.05 μm. It is clear from the plot in Fig. 4.11 that a much higher FSP would be expected if the FSP were measured at Griffin's value of water potential of -1 atm than at -30 atm. Therefore the high values of FSP obtained by Stone and Scallan and Griffin can be explained on this basis.

An explanation is also required for the high values of FSP determined by Feist and Tarkow (1967) and Ahlgren et al. (1972) by the polymer-exclusion technique, Stone and Scallan (1967) by the nonsolvent-water method, and Ahlgren et al. by the microscopic measurement of the cell-wall fraction of swollen wood. None of these methods involve the application of very low water potentials to thick wood specimens but they have one fact in common; they were all performed with microtome sections or pulped fibers. In such cases the restraint resulting from the cellular structure of wood is absent and, under such conditions, higher values of

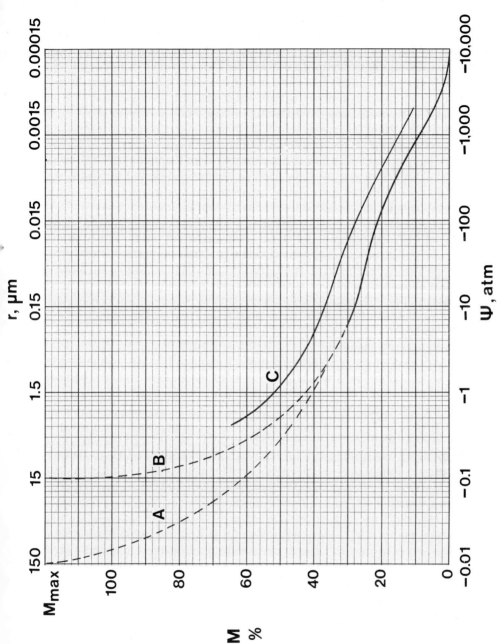

Fig. 4.11. Idealized moisture characteristic of wood: *A* hardwood with earlywood vessels. r = 150 μm; *B* softwood with tracheids with radii of 15 μm; *C Picea mariana* (Griffin 1977). The portion of curves *A* and *B* below $\psi = -25$ atm were calculated from Eq. (4.9) using the sorption-isotherm data in the Wood Handbook (USDA 1955)

EMC would be expected. It is also true that thin sections were used in the pressure-plate method. The curve obtained by Stone and Scallan for *Picea mariana* is also presented in Fig. 4.11. When this is compared with the idealized curve, the EMC values are significantly higher in the low-relative-humidity range when compared with EMC values based upon the Wood Handbook data. Again, this may be due to the lack of mechanical restraint in thin wood sections.

This discussion concerning the fiber saturation point of wood may be concluded with the observation that the value of FSP which is determined for a given wood specimen is dependent on the method used. The sorption isotherm of wood may be plotted using a linear scale of relative humidity as in Fig. 1.6 and, in this form is very useful for determination of the EMC in unsaturated air and in relating the moisture content to changes in physical properties, particularly when the FSP is accepted as the value extrapolated from 98% to 100%. However, as relative humidities are raised above 98%, corresponding to low water potentials, the sorption isotherm represents a continuum of values which approach full saturation of the wood at 100% relative humidity corresponding to a water potential of zero. The plot of the water characteristic of wood, consisting of EMC vs. log ψ, has a point of inflection at approximately -1 atm. If this point is taken as the FSP, the moisture content will be significantly higher than that corresponding with changes in the physical properties. It is advantageous then, from a practical standpoint, to define fiber saturation point as that moisture content corresponding to abrupt changes in physical properties.

Another aspect of the equilibrium between wood and its environment is that between wood and soil with which it is in contact. Leyton (1975) has published moisture characteristics for Chino clay and Indio loam soils, reproduced in Fig. 4.12. The permanent wilting points, corresponding to $\psi = -15$ atm occur at moisture contents of 15% for Chino clay and 5% for Indio loam. When wood is placed in contact with soil, equilibrium is expected to occur when the water potentials are equal. By combining Figs. 4.11 and 4.12, sorption isotherm plots of wood in the two soils may be obtained as represented in Fig. 4.13. In this case the permanent wilting point results in EMC's of approximately 25%, which is close to the assumed nominal FSP. As soil-moisture content increases beyond this point there is a very rapid rise in wood-moisture content toward full saturation of the wood at approximately 42% moisture content for Indio loam and 48% for Chino clay. Thus even in contact with the driest of soils, wood will reach moisture contents near 30%, which exceeds the 20% required to support fungal decay.

Baines and Levy (1979) have investigated the wick action of wood on water in which wood specimens were immersed within a closed container and exposed to air at the other end. The water potential difference causes movement of the water into the wood and from the wood to the air. A similar study was performed using soil (Baines 1980 unpublished results). In both these cases the results are in qualitative agreement with expected equilibrium between the water potentials in the wood and in water or soil, and between the wood and the unsaturated air.

The movement of sap in trees due to transpiration (Zimmermann 1983) can be explained by gradients of water potential. The water potential of the soil may be in the order of -1 atm, while that of unsaturated air may extend from $-1,500$ atm at h = 0.3 to -130 atm at h = 0.9, indicating a large available gradient between

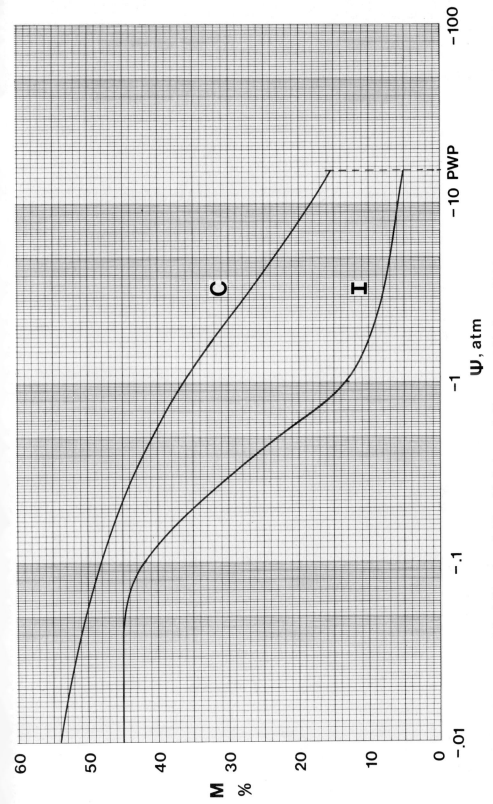

Fig. 4.12. Moisture characteristic of China clay soil (*C*) and Indio loam soil (*I*) as presented by Leyton (1975)

130 Capillarity and Water Potential

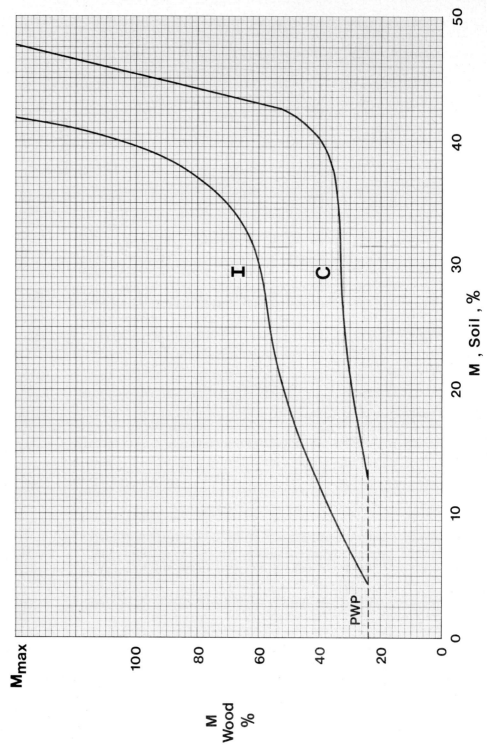

Fig. 4.13. Idealized sorption isotherm of wood in Indio loam (*I*) and China clay (*C*) soils, derived from Figs. 4.11 and 4.12

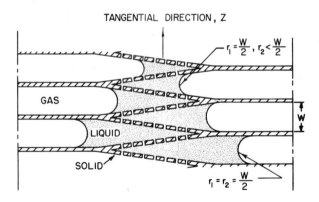

Fig. 4.14. Illustrating how capillary forces can cause free-water movement in the tangential direction of softwoods in accordance with the Comstock model as presented by Spolek and Plumb (1981). (Courtesy of Wood Sci. Technol.)

these two points. There must also be sufficient capillary tension to account for the height of the stem and viscous losses in the sap stream. Moisture can move from the soil to the roots due to the lower osmotic pressure within the roots. At this point the water potential may be -6 atm. If the stem is 300 feet high, the gravitational component must be -9 atm. In addition there are viscous losses through the cell lumens and pit openings and roots which may require an additional potential of -10 atm. Adding these together results in a tension at the mesophyll cells of approximately 26 atm. In order to account for such a tension, it is clear that the effective radius of the opening at the mesophyll cell must be less than 0.06 µm at the evaporating surface. If these assumptions are correct, then a water potential exceeding -26 atm in the moist air (corresponding to $h=0.98$) will result in moisture going into the tree rather than leaving it by transpiration. Such a reversal of movement is unlikely but could occur when the relative humidity exceeds 98%.

Spolek and Plumb (1981) show how the structure of softwoods can result in a gradient of matric water potential which can aid in the removal of free water from wood during drying. The gradient results when the free water has been removed from the straight-sided, cylindrical portions of the lumens and remains in the tapered ends. Since drying is most rapid at the outside surface, there is less free water in the tapered ends nearer the surface. As Fig. 4.14 shows, a decrease in moisture content toward the surface results in a decreased capillary radius and, therefore, a more negative water potential. Thus there is a gradient from the interior of the wood tending to force the free water out. It is clear from the figure, based upon the Comstock softwood model, that the direction of flow will be in the tangential direction because the tracheids are tapered along the radial surfaces.

Chapter 5

Thermal Conductivity

5.1 Fourier's Law

The steady-state flow of heat is described by Fourier's law, which is analogous to Darcy's law for the transport of liquids. The thermal conductivity of a material is equal to the flux divided by the gradient. Mathematically it may be stated as

$$K = \frac{H/tA}{\Delta T/L} = \frac{HL}{tA\Delta T} \qquad (5.1)$$

or as

$$H = \frac{KAt\Delta T}{L}, \qquad (5.2)$$

where K = thermal conductivity, cal/(cm °C s), H = quantity of heat transferred, cal, t = time interval, s, A = cross-sectional area perpendicular to direction of flow, cm^2, L = length of flow path in the transfer medium, cm, ΔT = temperature differential between the heat-transfer surfaces separated by L, °C.

It is evident from Eq. (5.1) that thermal conductivity is numerically equal to the rate of heat flow (in cal/s) through a unit cube of material (1 cm on a side) between two opposite parallel surfaces having temperature difference of 1 °C. The significance of the terms in Eq. (5.1) is explained in Fig. 5.1. When thermal conductivity is expressed in the British engineering system in common use in the United States, its value is based upon the heat flow in BTU/h through a unit slab of material with an area of one square foot and a thickness of one inch. In the SI system the rate of heat flow is expressed in joules per second or watts through a unit cube 1 m on a side.

The thermal conductivities of some common materials are given in Table 5.1 in both the cgs system and the British engineering system with thickness expressed in inches and area in ft^2 in the latter. Values of conductivity and conductance may be converted to other systems using the factors listed below

1 cal/(cm °C s) = 418 W/m K = 2,903 BTU in/(ft^2 h °F)
1 BTU in/(ft^2 h °F) = 0.0833 BTU/(ft h °F) = 0.144 W/m K
   = 3.44×10^{-4} cal/(cm °C s)
1 W/m K = 6.94 BTU in/(ft^2 h °F) = 2.39×10^{-3} cal/(cm °C s).
1 ft^2 h °F/BTU = 1 R = 0.176 m^2 K/W = 0.176 R_{SI}
1 m^2 K/W = 1 R_{SI} = 5.68 ft^2 h °F/BTU = 5.68 R
1 U = 5.68 U_{SI}
1 U_{SI} = 0.176 U.

It is clear from Table 5.1 that wood is a relatively good insulator, especially perpendicular to the fiber axis where there is high resistance to flow due to the inter-

Fig. 5.1. Significance of terms in Fourier's law

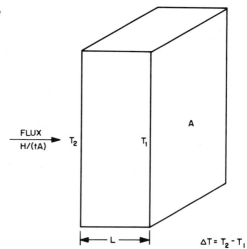

Table 5.1. Thermal conductivities

Material	K BTU in/(ft² h °F)	K 10⁴ cal/(cm °C s)
Wood, G=0.45 M=12% ⊥	0.87	3.0
Wood, G=0.45 M=12% ||	2.17	7.5
Wood, G=0.70 M=12% ⊥	1.23	4.25
Wood, G=0.70 M=12% ||	3.08	10.6
Cell-wall substance ⊥	3.05	10.5
Cell-wall substance ||	6.09	21.0
Water, free	4.80	14.0
Air, dead	0.165	0.57
Douglas-fir plywood	0.8	2.8
Hardboard, med. density	0.73	2.5
Particleboard, med. density	0.93	3.2
Asbestos-cement board	4.0	13.8
Cement mortar, common brick	5.0	17.2
Concrete	6.5	22.3
Stone, lime, sand	12.50	43.0
Glass	7.3	25.1
Fiberglass	0.27	0.93
Expanded polyurethane	0.16	0.55
Silver	2,900	10,000
Copper	2,680	9,230
Aluminium	1,400	4,820
Stainless steel	113	389

ruption of the path by the poorly conducting air-filled lumens. The insulating properties of wood have numerous advantages and contribute greatly to the fire-resistant properties of wood as compared with highly conducting metals, which soften at high temperatures. The low conductivity of wood also accounts for the considerable time interval required to bring a large-diameter pole or bolt to a uniform,

5.2 Empirical Equations for Thermal Conductivity

MacLean (1941) measured the thermal conductivities of many samples of wood with a large range of moisture contents and specific gravities. He presented an empirical equation which gave the best agreement with his experimental data:

$$K_{gT} = [G(5.18 + 0.096\ M) + 0.57\ v_a] \times 10^{-4}\ \text{cal}/(\text{cm}\ °\text{C}\ \text{s}), \tag{5.3}$$

(MacLean equation, M < 40%)
where K_{gT} = transverse thermal conductivity of wood, v_a porosity.

At moisture contents above 40%, a significant portion of the moisture is free water in the lumens, which apparently contributes more than the bound water to the over-all thermal conductivity. Therefore, the factor 0.096 in Eq. (5.3) is replaced by 0.131:

$$K_{gT} = [G(5.18 + 0.131\ M) + 0.57\ v_a] \times 10^{-4}\ \text{cal}/(\text{cm}\ °\text{C}\ \text{s}). \tag{5.4}$$

(MacLean equation, M > 40%)
When Eq. (1.19) is substituted into Eq. (5.3) its form may be simplified to

$$K_{gT} = [G(4.80 + 0.090\ M) + 0.57] \times 10^{-4}\ \text{cal}/(\text{cm}\ °\text{C}\ \text{s}) \tag{5.5a}$$

or

$$K_{gT} = [G(1.39 + 0.026\ M) + 0.165]\text{BTU in}/(\text{ft}^2\ \text{h}\ °\text{F}). \tag{5.5b}$$

(MacLean equation, M < 40%)
For higher moisture contents,

$$K_{gT} = [G(4.80 + 0.125\ M) + 0.57] \times 10^{-4}\ \text{cal}/(\text{cm}\ °\text{C}\ \text{s}) \tag{5.6a}$$

or

$$K_{gT} = [G(1.39 + 0.036\ M) + 0.165]\text{BTU in}/(\text{ft}^2\ \text{h}\ °\text{F}). \tag{5.6b}$$

(MacLean equation, M > 40%)
MacLean (1952) states that the ratio of longitudinal to transverse thermal conductivity varies from 2–1/4 to 2–3/4, with an average of approximately 2–1/2. Therefore it is assumed that

$$K_{gL} = 2.5\ K_{gT}, \tag{5.7}$$

where K_{gL} = longitudinal thermal conductivity of wood.

Equation (5.3) may be considered to represent a model consisting of parallel conductive paths of cell-wall substance, bound water, and air as represented in Fig. 5.2. Equation (5.5) predicts values of 7.6×10^{-4}, 9.6×10^{-4}, and 0.57×10^{-4} cal/(cm °C s) for dry cell-water substance, bound water, and air, respectively.

Fig. 5.2. Geometrical model which is representative of Eq. (5.3)

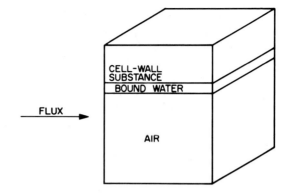

Equation (5.6) predicts a value of 13.1×10^{-4} cal/(cm °C s) for a mixture of bound and free water.

Maku (1954) calculated values of 10.0×10^{-4} and 15.6×10^{-4} cal/(cm °C s) for cell-wall substance perpendicular to and parallel with the fiber axis. Maku's calculations indicated that good agreement was obtained with data from the literature for wood when a value of 1.1×10^{-4} cal/(cm °C s) was assigned to air in the lumens. Kollmann and Malmquist (1956) determined a value of 9.7×10^{-4} cal/(cm °C s) for dry cell-wall substance perpendicular to the fiber axis.

5.3 Conductivity Model

In most transport processes, except hydrodynamic fluid flow, there is a significant flux through the cell-wall substance and usually, but not always, through the air in the lumens. Frequently, the flow through the pit openings may be neglected due to their small fractional surface area of the cell wall, which is usually less then 1% (see Chap. 6.6). A model is proposed similar to that of Fig. 1.9 except that the cell is assumed to have unit overall dimensions as indicated in Fig. 5.3. The model is assumed to apply to all longitudinal wood cells, including tracheids, fibers, vessels, and longitudinal parenchyma. The tangential and radial cell walls are assumed equal in thickness and the proportion of cell-wall thickness to diameter is assumed equal for all cells in a given wood specimen. The presence of walls at the ends of the cells, pit openings, and transversely oriented cells is neglected. This model is applicable to thermal and electrical conductivity, dielectric behavior, and water-vapor diffusion. A similar model with three components was proposed by Siau et al. (1968) for wood-polymer composite materials.

The mathematical analysis of the model with the flux in the transverse direction may be clarified by reference to Fig. 5.4. The cross walls are combined in Sect. (1) with a full length in the flow direction of $(1-a)$ and with a width equal to the unit width of the model. The side walls (3) are those portions of the cell wall aligned parallel with the flux and with a length equal to that of a lumen, a, and a combined width of $(1-a)$. The lumen (2) is a square having the dimension, a, on a side. The

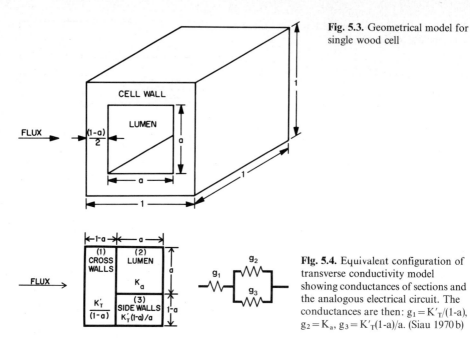

Fig. 5.3. Geometrical model for single wood cell

Fig. 5.4. Equivalent configuration of transverse conductivity model showing conductances of sections and the analogous electrical circuit. The conductances are then: $g_1 = K'_T/(1-a)$, $g_2 = K_a$, $g_3 = K'_T(1-a)/a$. (Siau 1970 b)

same flux flows through Sect. (1), as in Sects. (2) and (3), in parallel. Misalignment of cells and the flow across the boundary between Sects. (2) and (3) are neglected. Thus, the model is analyzed in a manner analogous to a series-parallel electrical circuit. Stamm (1964), Maku (1954) and Choong (1965) have discussed similar models, but with the flux at 90° to the direction indicated in Fig. 5.4.

It is clear from this discussion of the model that the parameter, a, is equal to the square root of the porosity of the wood

$$a = \sqrt{v_a} \tag{5.8a}$$

or

$$v_a = a^2. \tag{5.8b}$$

The porosity of wood may be calculated from the specific gravity and moisture content by means of Eqs. (1.18) or (1.19).

5.4 Resistance and Resistivity; Conductance and Conductivity

Resistance will be introduced as it applies to electrical circuits, which are directly analogous to thermal circuits. The resistance of a conductor is directly proportional to its length in the flow direction and inversely proportional to its cross-sectional area perpendicular to the flow direction. Expressed mathematically

$$R = r\, L/A, \tag{5.9}$$

where R=resistance, ohm, volt/ampere or s °C/cal, r=resistivity, ohm cm or cm °C s/cal, L=length, cm, A=cross-sectional area, cm².

Note that the resistivity is numerically equal to the resistance of a unit cube as measured between two parallel faces.

Equation (5.9) may be transformed to

$$g = K\, A/L, \tag{5.10}$$

where g=conductance, ohm^{-1} or cal/(s °C), K=conductivity, ohm^{-1} cm^{-1} or cal/(cm °C s).

It is clear, then, that conductance is the reciprocal of resistance and conductivity is the reciprocal of resistivity. Note the close resemblance between Eqs. (5.10) and (5.2). Heat conductance is then equal to $H/t\, \Delta T$.

5.5 Derivation of Theoretical Transverse Conductivity Equation

Electrical or thermal circuits must be analyzed in terms of resistances or conductances because resistivities and conductivities are specifically applicable to unit cubes of material. The resistance of the series-parallel circuits of Fig. 5.4 is then

$$R_T = R_1 + \frac{1}{1/R_2 + 1/R_3}, \tag{5.11}$$

where R_T=resistance of wood in the transverse direction, ohm or (s °C)/cal, R_1, R_2, R_3=resistances of Sects. (1), (2), and (3).

Equation (5.11) may be rewritten in conductance form as

$$\frac{1}{g_T} = \frac{1}{g_1} + \frac{1}{g_2 + g_3}. \tag{5.12}$$

Equation (5.10) may be used to determine the conductances of the sections in Fig. 5.4. Assuming a cubical model with unit dimensions,

$$g_T = K_{gT}$$

$$g_1 = \frac{K'_T}{(1-a)}$$

$$g_2 = K_a$$

$$g_3 = \frac{K'_T(1-a)}{a},$$

where K_{gT}=transverse thermal conductivity of single wood cell (or many such cells in combination), cal/(cm °C s), K_a=thermal conductivity of air, K'_T=transverse thermal conductivity of cell-wall substance.

When the above conductance values are substituted into Eq. (5.12), the following general conductivity equation is obtained:

$$\frac{1}{K_{gT}} = \frac{(1-a)}{K'_T} + \frac{a}{(1-a)K'_T + aK_a} \tag{5.13}$$

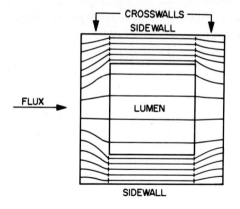

Fig. 5.5. Approximate relative flux concentrations through the cross walls, lumens, and side walls of the transverse model. The entire width of the cross walls is not fully effective for conduction due to the nonuniformity of the flux

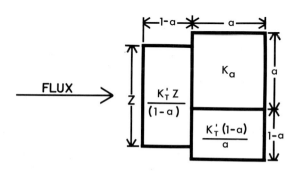

Fig. 5.6. Transverse conductivity model after modification for reduced conductive efficiency of the cross walls. (Siau 1970b)

or

$$K_{gT} = \frac{(1-a)K_T'^2 + aK_aK_T'}{(1-a)^2 K_T' + a(1-a)K_a + aK_T'} \tag{5.14}$$

where a = ratio of (lumen width)/(cell width) which, according to the assumptions of the model, is equal to the square root of porosity.

Figure 5.5 indicates how the flux concentration is nonuniform in the side walls, although it was assumed to be uniform in the derivation of Eq. (5.14). Since the entire width of cross wall is not effective for conduction because of this nonuniformity, a factor, Z, is introduced into Eq. (5.14) to account for this deficiency.

$$K_{gT} = \frac{(1-a)ZK_T'^2 + aZK_aK_T'}{(1-a)^2 K_T' + a(1-a)K_a + aZK_T'} \tag{5.15}$$

$M < M_f$

where Z = fraction of total cross-wall width which may be considered effective for conduction.

Equation (5.15) applies to a modified model as depicted in Fig. 5.6.

Hart (1964) has investigated the fringing effect on the current in the cross wall of the model, assuming no conductivity for the air space. He derived an empirical equation for calculating the fraction (w_1) of the cross-wall width adjacent to the lumen which may be considered effective for conduction, in which

$$w_1 = \frac{0.48(1-a)}{a}\{1-\exp[-2.08a/(1-a)]\},\tag{5.16}$$

where w_1 = fraction of that portion of the cross wall, located adjacent to the lumen with width a, which is effective for conduction, assuming zero conductivity for the lumen.

From the definition of w_1, it is clear that, with the assumption of zero flux in the lumen, the portion of cross-wall width adjacent to the lumen which is considered totally effective for conduction divided by the total cross-wall width of 1 is equal to w_1 a. The remaining portions of the cross wall adjacent to the side walls, with a total width of (1–a), are assumed to be fully conducting. In the case of heat flow, there is a significant flux in the lumen because of the conductivity of the air. Since the flux concentration in the lumen is (K_a/K_T') times that in the side walls, the additional fraction of cross-wall width totally effective for conduction may be calculated as (a K_a/K_T'). Therefore the total of these fractions of effective cross-wall width is w_1 a + (1–a) + a K_a/K_T', and this total fraction is defined as Z. Simplifying,

$$Z = 1 - a(1 - w_1 - K_a/K_T').\tag{5.17}$$

The value of Z as a function of a may by determined from Fig. 5.7.

Since Eq. (5.15) expresses K_{gT} as a function of a, it is useful to relate this latter variable to the specific gravity and to the density of the wood. The value of a is obtained from the specific gravity using Eq. (1.18) or (1.19), and the relationship between density and specific gravity is expressed by Eq. (1.17). These functions are plotted in Fig. 5.8. The relationship between porosity and specific gravity is depicted in Fig. 5.9.

The transverse thermal conductivity data from the literature are presented in Fig. 5.10. The values from the literature were converted to a fractional-cell-wall basis by use of Eq. (1.18).

The data of Wangaard (1943) were determined from his alignment chart for values of M of 0%, 12%, and 24%. These lines are very closely coincident. The results of Kollmann and Malmquist (1956) and Hendricks (1962) are in close agreement with Wangaard's data. Equation (5.3) of MacLean was also plotted for the same values of M, with the higher moisture contents yielding somewhat higher conductivities. Since most of these results except MacLean's show that the change in thermal conductivity with moisture content is small when calculated on a volume-fraction-of-cell-wall basis, there is a strong indication that the thermal conductivity of cell-wall substance (K_T') is independent of the moisture content and that it may be considered so for practical purposes. Theoretical points based on Eq. (5.15) are plotted for comparison with the literature values. Fairly good agreement is obtained when values of 1.0×10^{-4} and 10.5×10^{-4} cal/(cm °C s) are assumed for K_a and K_T', respectively, with the latter value being considered independent of moisture content. It is clear, however, that the value of K_{gT}, as calculated from Eq. (5.15), is not independent of moisture content because an increase in moisture content decreases the porosity of wood (and a), resulting in an increase in K_{gT}.

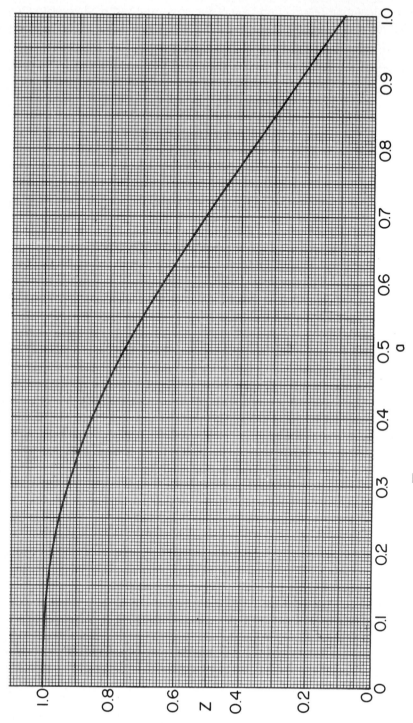

Fig. 5.7. Parameter Z as a function of a or $\sqrt{v_a}$

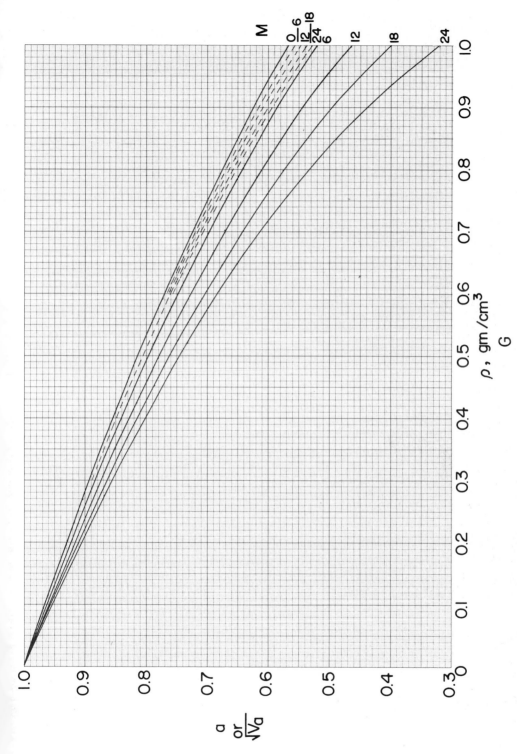

Fig. 5.8. Parameter a as a function of specific gravity and density at various moisture contents. The solid lines represent specific gravity and the dotted lines, density

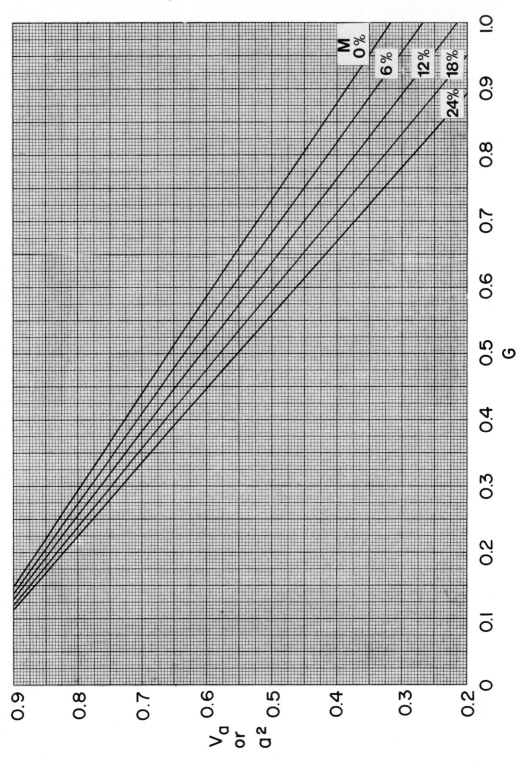

Fig. 5.9. Porosity as a function of wood specific gravity at various moisture contents according to Eq. (1.18)

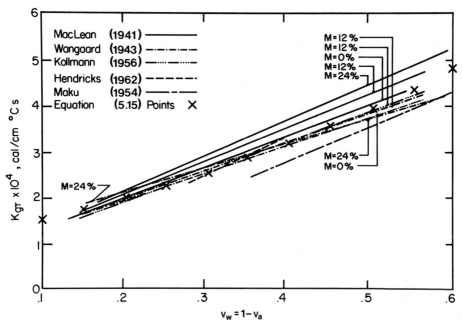

Fig. 5.10. Transverse thermal conductivities from the literature, as a function of the volume fraction of cell-wall substance. (Siau 1970b)

Equation (5.15) is plotted in Fig. 5.11 over the entire range of values of v_w and v_a, where $v_w = 1 - v_a$. This shows a nearly linear function up to v_w of 0.5, with increasing slope as K_{gT} approaches 10.5×10^{-4} cal/(cm °C s) for solid cell-wall substance.

When the values of 10.5×10^{-4} and 1.0×10^{-4} cal/(cm °C s) are substituted in Eq. (5.15), the following result is obtained:

$$K_{gT} = \frac{Z(110 - 99.5\,a)}{10.5 - a(20 - 10.5\,Z) + 9.5\,a^2} \times 10^{-4} \text{ cal/(cm °C s)}. \tag{5.18}$$

$M < M_f$

Equations (5.15) or (5.18) are plotted with a as the independent variable in Fig. 5.12. It is clear that this function is very nearly a straight line between values of a from 0.95 to 0.225 (G from 0.15 to 1.4). Since this covers almost all wood specific gravities, a useful regression equation may be obtained which has a maximum deviation of 2.8% from Eq. (5.15) within the above limits of G.

$$K_{gT} = (12.2 - 11.3\,a) \times 10^{-4} \text{ cal/(cm °C s)}. \tag{5.19}$$

Equation (5.19) is plotted in Fig. 5.12 for comparison with plot of Eq. (5.15).

5.6 Derivation of Theoretical Longitudinal Conductivity Equation

When the model of Fig. 5.3 is applied with the flux in the direction of the fiber axis, the following theoretical equation results:

$$K_{gL} = K'_L (1 - v_a) + K_a v_a \tag{5.20a}$$

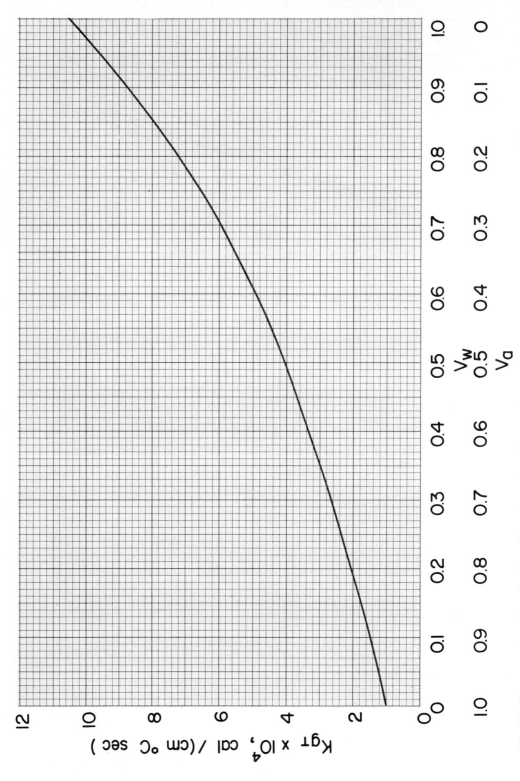

Fig. 5.11. Transverse thermal conductivity of wood vs. volume fraction of cell-wall substance and porosity according to Eq. (5.15)

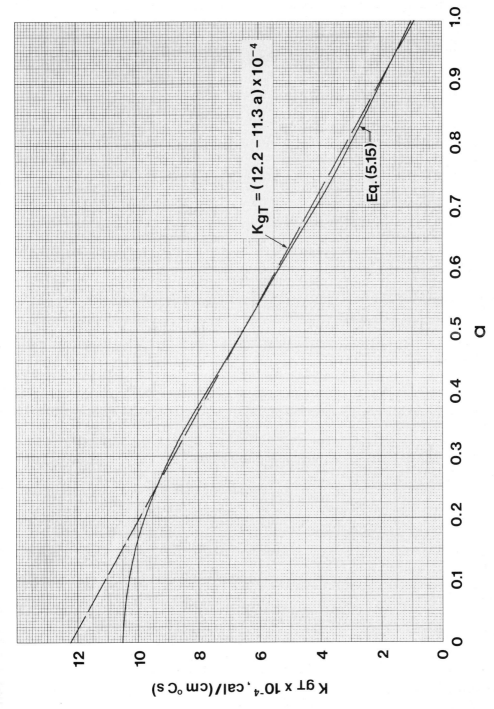

Fig. 5.12. Transverse thermal conductivity of wood vs. parameter a, according to Eq. (5.15) and the linear regression line of Eq. (5.19)

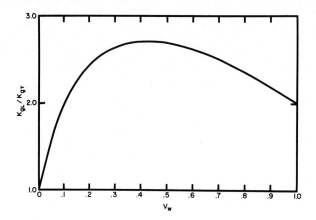

Fig. 5.13. Ratio of longitudinal to transverse thermal conductivities as a function of the volume fraction of cell-wall substance. (Siau 1970b)

or

$$K_{gL} = K'_L (1-a^2) + K_a a^2, \quad (5.20\,\text{b})$$

where K'_L = longitudinal thermal conductivity of cell-wall substance.

A value of 21.0×10^{-4} cal/(cm °C s) for K'_L results in ratios of K_{gL}/K_{gT} as shown in Fig. 5.13. These are between the values of 2.25 and 2.75 for the usual wood specific gravities in agreement with the findings of MacLean (1952).

When the above value is substituted into Eq. (5.20), it has the form

$$K_{gL} = (21 - 20\, v_a) \times 10^{-4} \text{ cal/(cm °C s)} \quad (5.21\,\text{a})$$

or

$$K_{gL} = (21 - 20\, a^2) \times 10^{-4} \text{ cal/(cm °C s)}. \quad (5.21\,\text{b})$$

5.7 R and U Values; Convection and Radiation

The R-value represents the thermal resistance of a path with unit area and therefore is directly proportional to the length in the direction of heat flow. It may be calculated by dividing this length by the thermal conductivity which is representative of a unit cube or slab.

$$R = L/K \quad (5.22\,\text{a})$$

or

$$R_{SI} = L/K, \quad (5.22\,\text{b})$$

where R = thermal resistance, ft² h °F/BTU, R_{SI} = thermal resistance, m² K/W, L = thickness, in or m.

The R-value is particularly useful in series thermal paths composed of several layers through which the heat flux passes. As in electrical circuits, it may be stated that

$$R_t = R_1 + R_2 \ldots + R_i, \quad (5.23)$$

where R_t = total resistance of a series consisting of $R_1, R_2 \ldots R_i$.

Household insulations available in North America are rated by R-value in the British Engineering System. When these are applied inside a wall which may also include wood siding and sheathing on the outside and a panel product on the interior, the total thermal resistance is equal to the total of the components. A wood with a specific gravity of 0.45 and a moisture content of 12% will have a thermal conductivity of approximately 0.87 BTU in/(ft² h °F). Therefore this has an R-value of 1.15 per inch. A 3-inch thickness will be R-3.45. In order to determine the rate of heat flow through this material it is more convenient to use the conductance or *U-Value* which is the reciprocal of the resistance.

$$U = 1/R = K/L$$
$$U_{SI} = 1/R_{SI} = K/L, \qquad (5.24)$$

where U = conductance, BTU/(h ft² °F), U_{SI} = conductance, W/m² K.

The rate of heat flow may be determined by substitution into Eq. (5.1).

$$H/t = U \ A \ \Delta T = U_{SI} \ A \Delta T, \qquad (5.25)$$

where H/t is expressed in BTU/h or W.

The use of thermal conductance values facilitates calculations involving parallel thermal paths. In such cases, each parallel path may have a different conductance, but an average conductance may be determined from a sum of the conductances and area fractions of the components.

$$\bar{U} = U_1 a_1 + U_2 a_2 \ldots + U_i a_i, \qquad (5.26)$$

where $\bar{U}$ = average conductance of a total area, $U_1, a_1 \ldots U_i, a_i$ = conductances and area fractions of the components such that $\sum_{0}^{i} a_i = 1$.

Heat is conducted to or from a wall or surface by a combination of *convection* and *radiation*. The mechanics of the convection process results in the formation of a *viscous* or *turbulent* boundary layer at the surface which may be characterized by a film coefficient (h) which has the same units as conductance. Therefore its reciprocal (1/h) is a resistance which may be added to the thermal resistance of the wall. Film coefficients depend on many factors: the temperature difference between the surface and the bulk of the surrounding air, the velocity of the bulk air, the viscosity, density, and thermal diffusivity of the air. These properties are accounted for by dimensionless groups such as Reynolds' number (dimensionless velocity), Grashoff number (ratio of gravitational forces to viscous drag), and the Prandtl number (ratio of kinematic viscosity to thermal diffusivity). These may be combined in various ways to calculate the Nusselt number, which may be regarded as a dimensionless convective film coefficient. The latter is the ratio of the convective film coefficient to the thermal conductance along the surface under consideration. The determination of a coefficient also requires knowledge of whether the convection is *free* (natural) or *forced*, either of which may have a viscous or turbulent boundary layer. The critical Reynolds' number for a boundary layer over a surface is 20,000. Reynolds' number for a plane surface is

$$Re = \frac{L \bar{v} \varrho}{\eta}, \qquad (5.27)$$

Table 5.2. Conductances of air films and air spaces (ASHRAE 1977)

	Conductance, BTU/(h ft² °F)
Inside wall surfaces, natural convection, laminar boundary layer, h_i	1.47
Exterior surface, forced convection, turbulent boundary layer, 15 mph wind, h_o	5.88
Exterior surface, 7 1/2 mph wind, h_o	4.0
Nonreflective air space, vertical, over 3/4" in thick, $\bar{T} = 50$ °F, $\Delta T = 10$ °F	0.99
Nonreflective air space, horizontal, $\bar{T} = 50$ °F, $\Delta T = 10$ °F	0.9

where
L = effective length parallel to the surface along which the boundary layer is formed,
$\bar{v}$ = average velocity of bulk air relative to the surface.

Details regarding the calculation of convective film coefficients are available in Leyton (1975) and the ASHRAE Handbook (ASHRAE 1977). The latter reference includes an approximate coefficient for radiation of 0.7 BTU/ft² h °F).

Combined coefficients which include the effects of convection and radiation heat transfer are of more practical value for the calculation of heat losses from a building. On an interior surface, the air is essentially stationary. Therefore free convection with a viscous boundary layer occurs with a relatively low conductance. The ASHRAE Handbook (ASHRAE 1977) gives a value of 1.47 BTU/(h ft² °F) for h_i, the surface coefficient for an interior wall, corresponding to an R-value of 1/1.47 = 0.61. On the exterior surface there is forced convection in the turbulent mode, resulting in a much higher coefficient which is affected principally by wind velocity. In this case the ASHRAE Handbook suggests a value of 5.88 BTU/(h ft² °F) in a 15 mile-per-hour wind, and 4.0 in a 7 ½ mile-per-hour wind, for h_o, the exterior surface coefficient. These values of h_o correspond to R-values of 0.17 and 0.25 respectively.

An *example* will illustrate the determination of the total effective resistance of the wall of a building. Assume a wall is made of solid timbers 8 inches thick of a wood with G = 0.45 and M = 12%. Then

$$\text{Effective R} = \frac{1}{h_i} + \frac{L}{K_{gT}} + \frac{1}{h_o}. \tag{5.28}$$

Substituting numerical values, $1/1.47 + 8/0.87 + 1/5.88 = 0.61 + 9.20 + 0.17 = 9.98$ ft² h °F/BTU. In this particular case the total contributions of convection and radiation are R-0.78, or 7.8% of the total. At lower wind velocities, the resistance of the outside surface will increase significantly, approaching the value for the inside surface in still air.

In order to calculate the rate of heat loss, the effective U = 1/9.98 = 0.1002 BTU/(ft² h °F) is multiplied by the total area and the temperature difference.

The conductance values discussed above are summarized in Table 5.2. In regard to wide air spaces (over 3/4 in) the conductance is nearly independent of thickness; therefore a thin air space has nearly the same resistance as a thick one due to the effect of convection currents. It is also clear that the effective conductivity of the air increases as thickness increases from a value of 0.99 (0.75) = 0.74 BTU in/(ft² h °F) at L = 0.75 in to 0.99 (3.5) = 3.47 at L = 3.5 in. Even at L = 0.75 in, the conductivity is much higher than 0.165, the value for dead air.

5.8 Application to Electrical Resistivity Calculations

Since the resistance of the air in the lumens is assumed to be infinite, the value of Z may be calculated as

$$Z = 1 - a(1 - w_1). \qquad (5.29)$$

The electrical resistivity analog of Eq. (5.15) for resistivity in the transverse direction is

$$r_T = r'_T \left[\frac{1-a}{Z} + \frac{a}{1-a} \right], \qquad (5.30)$$

where r_T = transverse resistivity of wood, ohm cm, r'_T = transverse resistivity of cell-wall substance, ohm cm.

The electrical resistance analog of Eq. (5.20) for resistivity in the longitudinal direction is

$$r_L = \frac{r'_L}{1 - a^2}, \qquad (5.31)$$

where r_L = longitudinal resistivity of wood, ohm cm, r'_L = longitudinal resistivity of cell-wall substance, ohm cm.

The model equations are not as useful for electrical calculations as they are to thermal conductivity problems because there are several factors which affect the electrical resistance of wood to a much greater extent than structural model considerations. For example, the resistance of wood decreases sharply with increases in temperature and moisture content, as discussed in detail by Davidson (1958), Brown et al. (1963), and Lin (1965). Also, the presence of small quantities of inorganic impurities in the cell wall can decrease its resistance significantly. All the above factors cause large variations in the resistivity of the cell-wall substance, while its thermal conductivity is essentially constant.

The electrical resistivity of wood is a useful measure of its moisture content. Values for various species are tabulated in the Wood Handbook (USDA 1955). Such data are also available in the instructions for electrical-resistance moisture meters. Generally, such meters are most effective from approximately 6% to 20%. The accuracy is relatively poor above the fiber saturation point.

5.9 Application to Dielectric Constant Calculations

Equation (5.15) may be used to calculate the dielectric constant of wood by direct substitution of the dielectric constant of cell-wall substance into the equation. Since the dielectric constant of air is 1.0,

$$\varepsilon_T = \frac{(1-a) Z \varepsilon'^2_T + a Z \varepsilon'_T}{(1-a)^2 \varepsilon'_T + a(1-a) + a Z \varepsilon'_T}, \qquad (5.32)$$

where ε_T = transverse dielectric constant of wood, ε'_T = transverse dielectric constant of cell-wall substance. The correction factor is then

$$Z = 1 - a(1 - w_1 - 1/\varepsilon'_T). \tag{5.33}$$

The dielectric analog of Eq. (5.20) may be written as

$$\varepsilon_L = \varepsilon'_L(1-a^2) + a^2, \tag{5.34}$$

where ε_L = longitudinal dielectric constant of wood, ε'_L = longitudinal dielectric constant of cell-wall substance.

Note that the values of ε'_T and ε'_L apply to moist cell-wall substance, and that these values change with moisture content and frequency, as explained by Skaar (1948).

The dielectric constant of wood is affected by moisture content, temperature, frequency, and fiber direction. Therefore, the use of a model equation is subject to limitations similar to those which apply to the electrical resistivity equation. The change in dielectric constant with moisture content is the basis of the dielectric or capacitance-type moisture meter.

Chapter 6

Steady-State Moisture Movement

6.1 Fick's First Law Under Isothermal Conditions

Diffusion is molecular mass flow under the influence of a concentration gradient. Therefore a static pressure difference is not necessary for diffusion to occur. Fick's first law, which is analogous to Darcy's and Fourier's laws, represents the relationship between the flux and the concentration gradient under steady-state conditions. When applied to water-vapor transport through wood is may be written as

$$D = \frac{w/tA}{\Delta c/L}, \tag{6.1}$$

where D = water-vapor diffusion coefficient of wood, cm^2/s, w = mass of water vapor transferred through wood in time t, g, A = cross-sectional area of specimen, cm^2, L = length in flow direction, cm, t = time, s, Δc = concentration difference, g/cm^3.

For convenience, the concentration difference can be expressed in terms of the moisture content difference as

$$\Delta c = \Delta M \, G \, \varrho_w / 100, \tag{6.2}$$

where G = specific gravity of wood at moisture content, M, ϱ_w = normal density of water = 1 g/cm^3.

Then, by substitution

$$D = \frac{100 \, wL}{tA\Delta M\varrho_w}. \tag{6.3}$$

There are several alternative ways of expressing the potential which drives moisture through wood. In addition to moisture content, there is the partial water vapor pressure in equilibrium with the wood, the relative humidity in equilibrium with the wood, the chemical or water potential, and the spreading pressure (Skaar and Babiak 1982). Skaar (1954) has shown that isothermal moisture transport calculations may be made using concentration, moisture-content, or partial-vapor pressure gradients with identical results. The alternative transport equations are written below in derivative form.

Transport Equation		Coefficient	Potential
$J = -D \, dc/dx$	(6.4a)	D, cm^2/s	c, gm/cm^3
$J = -K_M \, dM/dx$	(6.4b)	K_M, g/(cm s%)	M, %
$J = -K_p \, dp/dx$	(6.4c)	K_p, g cm/(dyne s)	p, dyne/cm^2
$J = -K_\mu \, d\mu/dx$	(6.4d)	K_μ, g mol/(cm s cal)	μ, cal/mol
$J = -K_\psi \, d\psi/dx$	(6.4e)	K_ψ, g/(cm atm s)	ψ, atm
$J = -K_\phi \, d\phi/dx$	(6.4f)	K_ϕ, g/(dyne s)	ϕ, dyne/cm

where J = flux, g/cm² s, p = partial vapor pressure, μ = chemical potential, ψ = water potential, ϕ = spreading pressure.

All of the coefficients listed above may be expressed in terms of D, which is the one most commonly used in the literature. For example, the coefficient $K_M = D \, \partial c / \partial M$. Since $\partial c = G \, \varrho_w \, \partial M / 100$,

$$K_M = \frac{G \varrho_w D}{100}. \tag{6.5}$$

Similarly, $K_p = D \, \partial c / \partial p$. Since $100 \, p/p_0 = H$, $\partial p = p_0 \, \partial H / 100$, and

$$K_p = \frac{G \varrho_w D}{p_0} \frac{\partial M}{\partial H}. \tag{6.6}$$

It is clear from Eq. (6.6) that the sorption isotherm of wood is needed to relate K_p to D.

The chemical potential of the bound water in wood may be calculated as

$$\mu = \mu_1^0 + RT \ln (H/100), \tag{6.7}$$

where μ = chemical potential, cal/mol, μ_1^0 = chemical potential of liquid water 1 atm and temperature, T, R = 1.987 cal/mol K.

Under isothermal conditions, μ_1^0 is a constant with a derivative of zero, then $d\mu = 100 \, RT \, dH/H$, and the coefficient for chemical potential can be calculated as

$$K_\mu = \frac{G \varrho_w H D}{100 \, RT} \frac{\partial M}{\partial H}. \tag{6.8}$$

Similar expressions can be derived for the calculation of K_ψ, and K_ϕ from D and the sorption isotherm of the wood.

It is clear from the above that the coefficient for the transport of water vapor through wood may be measured by applying gradients of moisture content, partial vapor pressure, or chemical potential. The values obtained can then be converted to D, which is based upon a bound-water concentration gradient with identical results in all cases.

6.2 Bound-Water Diffusion Coefficient of Cell-Wall Substance

Stamm (1959) has measured the longitudinal bound-water diffusion coefficient of the cell-wall substance of the wood of *Picea sitchensis* at 26.7 °C. This was done by filling the lumens with a low-melting alloy of bismuth, lead, and tin. The specimens had thicknesses between 0.1 and 0.2 inch. Measurements were made using the unsteady-state method (Chap. 7.3.1) by taking the slopes of the moisture content vs. square-root-of-time curves at the point of two-thirds saturation. Stamm's results are plotted in Fig. 6.1, where it is evident that there is an exponential increase with wood-moisture content. This may be explained by a lower bonding energy between the sorption sites and the bound-water molecules at higher moisture contents. This bonding energy should approach zero at the fiber saturation point where subsequently added moisture is essentially free water, held by relatively small capillary forces.

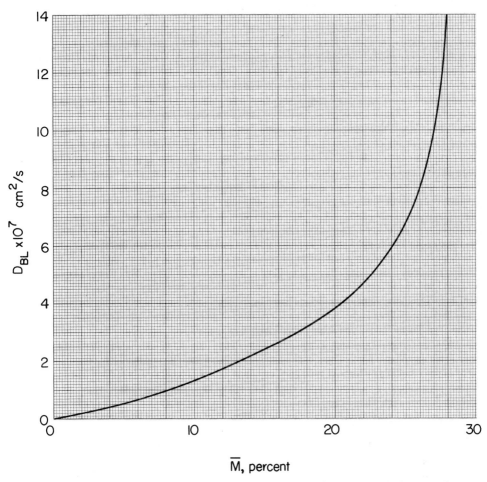

Fig. 6.1. Longitudinal bound-water diffusion coefficient (D_{BL}) of the cell-wall substance of *Picea sitchensis* at 26.7 °C as a function of the average moisture content. (According to Stamm 1959)

In the determination of the diffusion coefficient it was necessary to initially equilibrate the wood to a given relative humidity and then expose the specimen to a higher relative humidity or to liquid water. The data in Fig. 6.1 therefore represent values for adsorption. Comstock (1963) found lower values of diffusion coefficients when measured by adsorption than by desorption at moisture contents above 12% when measurements were made by both the steady-state (Sect. 6.9) and the unsteady-state methods. He also found higher values by the steady-state method. These effects were attributed to the decrease in the EMC of wood due to compressive stress. Since desorption is employed in the drying of wood, and since relatively high air velocities are usually employed, the values obtained by Stamm would be expected to be considerably lower than those existing during the drying of wood.

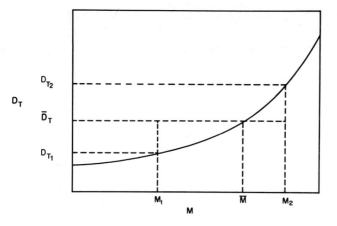

Fig. 6.2. Illustrating the relationship between diffusion coefficient and moisture content at constant temperature and the determination of the average diffusion coefficient and the corresponding average moisture content

Stamm's (1959) unsteady-state plots of relative saturation vs. square root of time were nearly linear up to a point of approximately two thirds of saturation. Since a range of moisture contents was therefore used in the determination of the diffusion coefficient, it was necessary to assign an average moisture content to the diffusion coefficient calculated from the plot. Stamm used the moisture content at the two-thirds saturation point calculated as

$$\bar{M} = M_1 + 2/3 \, (M_2 - M_1), \tag{6.9}$$

where $\bar{M}$ = average moisture content at which the average diffusivity is approximately equal to the true coefficient, M_2 = higher moisture content, M_1 = lower moisture content.

Since the diffusion coefficient is a function of moisture content, it is continuously variable throughout a specimen where there is a variation in moisture content. The moisture profile is not known, but it cannot be linear because D is not a constant. Since flux is a product of the coefficient and gradient, a relationship between the measured average diffusion coefficient and the true variable diffusion coefficient may be written as

$$\bar{D}_T (M_2 - M_1) = \int_{M_1}^{M_2} D_T dM . \tag{6.10}$$

The integral in Eq. (6.10) cannot be evaluated exactly unless the moisture profile across the specimen is known. Stamm (1959) indicates that moisture profiles tend to be parabolic. Figure 6.2 illustrates the relationship described by Eq. (6.10), according to which the area bounded by the curve between M_1 and M_2 is equal to the rectangular area $\bar{D}_T (M_2 - M_1)$. It is clear from the figure that this equality occurs at an average moisture content exceeding the arithmetic average. In the case of a parabolic moisture profile, this occurs at a moisture content approximately 60% of the way between the low and high moisture contents. This is therefore approximately equal to the value obtained from Eq. (6.9). This relationship is not applicable longitudinally in wood because the coefficient decreases with a rise in moisture content due to the contribution of conducting paths through the air in the lumens (Sect. 6.7).

Stamm (1964) found that the longitudinal bound-water diffusion coefficient of cell-wall substance was approximately three times that in the tangential and two times that in the radial directions. Assuming an average value of 2.5, the following equation may be used to calculate transverse values:

$$D_{BL} = 2.5\ D_{BT} \tag{6.11}$$

where D_{BL} = longitudinal bound-water diffusion coefficient of cell-wall substance, D_{BT} = transverse bound-water diffusion coefficient of cell-wall substance.

In the drying of wood, it is necessary to know the fiber saturation point to calculate diffusion coefficients and drying times. The value of 30% is assumed for domestic woods at 20 °C. This decreases approximately 0.1% per degree rise in temperature. Stamm and Nelson (1961) published values of FSP up to 120 °C as plotted in Fig. 6.3. Therefore, if a specimen were dried to 6% at 20 °C, the average moisture content for determining $\bar{D}_T$ would be 6 + 2/3 (30) = 26%. If this were done at 100 °C, it would be 6 + 2/3 (22.5) = 21%.

6.3 The Combined Effect of Moisture Content and Temperature on the Diffusion Coefficient of Cell-Wall Substance

It will be shown later that the principal resistance to bound-water diffusion through wood in the transverse direction is in the cell wall. Stamm (1964) and Choong (1963) have found that the coefficient increases rapidly with temperature in accordance with the Arrhenius equation, with an activation energy of approximately 8,500 cal/mol at a moisture content of 10%. The relationship may be written as

$$D_T = C\ \exp(-E_b/RT), \tag{6.12}$$

where D_T = transverse bound-water diffusion coefficient of wood, cm²/s, C = constant, E_b = activation energy = 8,500 cal/mol at M = 10%.

Since most of the resistance to transverse diffusion is in the cell wall, Eq. (6.12) will be assumed to apply also to cell-wall substance. When Stamm's values of D_{BL} in Fig. 6.1 at 26.7 °C are fitted to Eq. (6.12) with E_b assumed to be 8,500 cal/mol at M = 10%, the activation energies extend from 9,500 cal/mol at M = 2% to 7,100 cal/mol at M = 28%. The result is illustrated in Fig. 6.4. Although the function is curvilinear, it may be closely approximated by a straight line in the region of primary interest between 5% and 25%. Then activation energy as a function of moisture content may be calculated as

$$E_b = 9,200 - 70\ M\ \text{cal/mol}. \tag{6.13}$$

When this and Eq. (6.11) are substituted into Eq. (6.12) and the constant adjusted to the optimum value, Stamm's values for D_{BT} may be approximated as

$$D_{BT} = 0.07\ \exp[-(9,200 - 70\ M)/R\ T], \tag{6.14}$$

where M is between 5% and 25%.

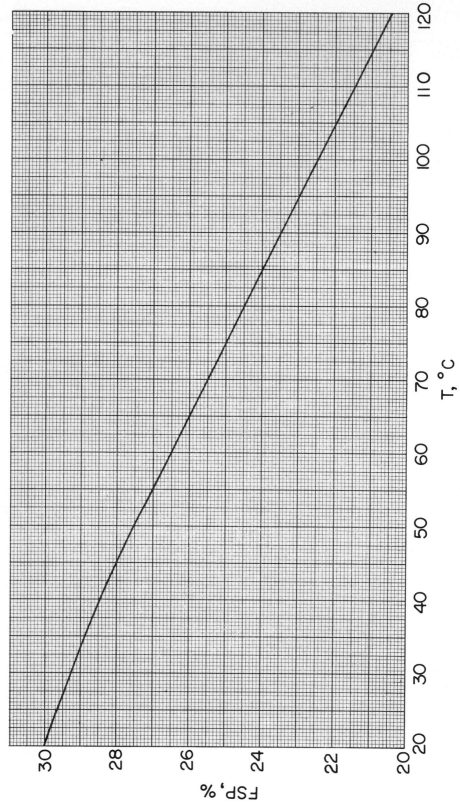

Fig. 6.3. Decrease in fiber saturation point with temperature. (According to Stamm and Nelson 1961)

Fig. 6.4. The molar heat of vaporization of bound water in wood (E_v) and the activation energy for bound-water diffusion (E_b) as calculated from the data of Stamm (1959) using the Arrhenius equation, assuming a value of 8,500 cal/mol at a moisture content of 10%. Also shown is a linear approximation as stated in Eq. (6.13)

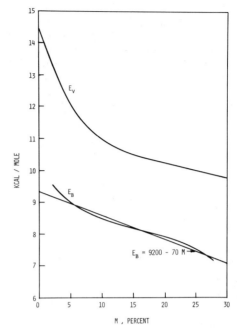

The values of D_{BT} obtained from Eq. (6.14) are plotted in Fig. 6.5 for 20 °C intervals of temperature to 100 °C and for moisture contents between 5% and 25%. The strong dependency of D_{BT} on both temperature and moisture content represents a significant deviation from Fick's law, which is based upon a constant coefficient. Although Fick's law is not obeyed, Eqs. (6.4a) through (6.4f) represent convenient relationships between flux and gradient, and their accuracy may be improved by calculation of the appropriate value of $\bar{M}$ from Eq. (6.9) or by employing an iterative solution in which values of the coefficient are continuously varied with the moisture content.

6.4 Water-Vapor Diffusion Coefficient of Air in the Lumens

In order to be able to calculate a theoretical diffusion coefficient for wood, it is necessary to know the coefficient for the transport of water vapor through the lumens. This may be calculated from the coefficient of interdiffusion of water vapor in air. A semi-empirical equation for this coefficient was derived by Dushman (1962)

$$D_a = 0.220 \left(\frac{76}{P}\right)\left(\frac{T}{273}\right)^{1.75}, \qquad (6.15)$$

where D_a = coefficient of interdiffusion of water vapor in bulk air, cm²/s, P = total pressure, cm Hg, T = Kelvin temperature.

In order to be useful in modeling diffusion through wood, D_a, which is based upon a moisture concentration gradient in air, must be converted to a basis of con-

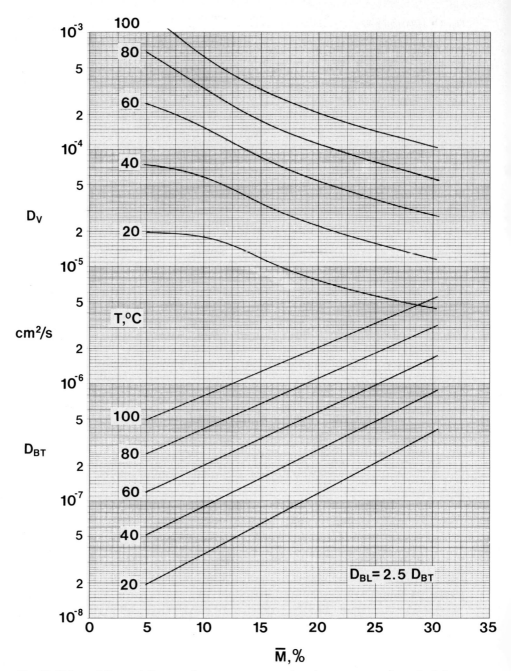

Fig. 6.5. Values of D_{BT} and D_v at various temperatures and moisture contents determined form Eqs. (6.14) and (6.18)

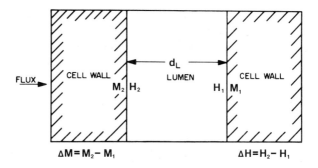

Fig. 6.6. Illustrating the corresponding moisture-concentration differences in the cell wall and in the lumen

centration of moisture in the cell-wall substance in equilibrium with the air. This requires a knowledge of the sorption isotherm because it is assumed that the water-vapor concentration in the lumens is in equilibrium with the cell-wall moisture content as depicted in Fig. 6.6. If the moisture is diffusing from left to right, $M_2 > M_1$ and $H_2 > H_1$. Therefore the relative humidity gradient in the lumen must correspond with a wood-moisture gradient as defined by the sorption isotherm. When D_a is converted to a gradient based on concentration in the cell wall, it is designated as D_v, the water-vapor diffusion coefficient of air in the lumens of wood. Both D_a and D_v are, by definition, equal to flux divided by gradient. The fluxes are the same for both and only the gradients are different, therefore the conversion requires a knowledge of the appropriate gradients. The gradient for D_a may be calculated from the relative humidity and saturated vapor pressure by means of the general gas law. In this specific case,

$$pV = \frac{w}{18} RT,$$

where p = partial water-vapor pressure in the lumen, w = mass of water vapor in volume, V.

Since $p = p_0 \times H/100$, the concentration of water vapor in the lumens is

$$\frac{w}{V} = \frac{18 \, p_0 H}{100 \, RT},$$

and the gradient is

$$\frac{18 \, p_0 \Delta H}{100 \, RT d_L} \, \frac{\text{g, moisture}}{\text{cm}^3 (\text{air}) \, \text{cm}}, \tag{6.16}$$

(Gradient in air)

where $\Delta H = H_2 - H_1$, d_L = diameter of lumen, 18 g/mol = molecular weight of water.

The corresponding gradient on a basis of concentration in cell-wall substance is a much larger quantity due to relatively high concentration of moisture in wood compared to that in air in equilibrium with it. Its value is

$$\frac{\Delta M G'_M \varrho_w}{100 \, d_L} \, \frac{\text{g, moisture}}{\text{cm}^3 (\text{cell wall}) \, \text{cm}}, \tag{6.17}$$

(Gradient in cell wall)

where G'_M = specific gravity of moist cell-wall substance, numerically equal to oven-dry weight divided by moist volume, ϱ_w = normal density of water, 1 g/cm^3.

The value of D_a may then be converted to D_V by multiplying by the concentration gradient in air [Eq. (6.16)] and dividing by the gradient in the cell wall [Eq. (6.17)]. Then,

$$D_V = \frac{18 \, D_a p_0}{G'_M \varrho_w RT} \frac{dH}{dM}, \tag{6.18}$$

where the inverse slope of the sorption isotherm is expressed as a derivative, $R = 6{,}230 \text{ cm}^3 \text{ cm Hg/mol K}$.

The value of G'_M may be calculated from the oven-dry value of 1.5.

$$G'_M = \frac{1.5}{1 + 0.015 \, M}. \tag{6.19}$$

Values of D_V calculated from Eq. (6.18) are plotted in Fig. 6.5 along with D_{BT}. It is clear from this that D_V is three to four orders of magnitude smaller than D_a because of the much greater concentration of moisture in the cell wall than in the air in equilibrium with it. Also D_V increases with M at lower moisture contents and temperatures and then decreases significantly in sharp contrast with the increase in D_{BT} with M. The reason for this is the point of inflection of the sorption isotherm where the slope dM/dH stops decreasing and begins to increase. The calculations are based upon the sorption data given in the Wood Handbook (USDA 1955). When applied to a specific wood specimen, sorption data determined for that specimen should be used in the determination of D_V.

6.5 The Transverse Moisture Diffusion Model

It is clear from Fig. 6.5 that both D_V and D_{BT} vary greatly with temperature and moisture content and these factors must be taken into account if it is desired to calculate the transverse diffusion coefficient of wood, D_T, from a model equation similar to Eq. (5.18) used for thermal conductivity. The thermal conductivities of the cell wall and air in the lumens are relatively constant and have a ratio of approximately 10 to 1. This relationship is reversed in the case of water-vapor diffusion with the ratio of D_{BT} to D_V ranging from 0.05 to 0.0005, corresponding to a ratio of D_V to D_{BT} from 20 to 2,000. This fact permits considerable simplification in the derivation of the model equation.

The transverse model is illustrated in Fig. 5.4. The lumen and side walls are in parallel and, in this combination, the very low conductance of the side walls, g_3, may be neglected. Considering the series combination of the conductance of the cross walls, g_1, and that of the lumen, g_2, the conductivity Eq. (5.12) may be simplified to the form

$$\frac{1}{g_T} = \frac{1}{g_1} + \frac{1}{g_2}, \tag{6.20}$$

where g_T = conductivity of wood, g_1 = conductivity of cross walls, g_2 = conductivity of lumens.

When the model is applied to moisture diffusion, the conductivities of the cell wall and air are replaced by the diffusion coefficients D_{BT} and D_v. Since these quantities are expressed on a basis of concentration in cell-wall substance, they both must be divided by the volume fraction of cell-wall substance in wood, represented by v_w, $(1-v_a)$, or $(1-a^2)$. Therefore the conductances g_1 and g_2 are derived similarly to those for heat flow (Chap. 5.5) except for the division by $(1-a^2)$. Then

$g_T = D_T$, $g_1 = D_{BT}/(1-a)(1-a^2)$, $g_2 = D_v/(1-a^2)$.

Substituting into Eq. (6.20) and solving for D_T,

$$D_T = \frac{1}{(1-a^2)} \frac{D_{BT} D_v}{D_{BT} + D_v(1-a)}, \quad (6.21)$$

$M < M_f$

where $a^2 = v_a =$ porosity of wood.

At moisture contents less than 15%, the conductance of the lumen (g_2) becomes sufficiently high relative to that of the cross walls (g_1) that it may be neglected with a maximum error of a few percent in the results. When this is done, the conductance of the wood is equal to g_1 and D_T may be calculated as

$$D_T = \frac{D_{BT}}{(1-a^2)(1-a)}. \quad (6.22)$$

$M < 15\%$

Equations (6.21) and (6.22) are based on the assumption of uniform flux in the cross walls. Actually there is a bending of flux toward the center of the lumens opposite to the bending of thermal flux illustrated in Fig. 5.5. This is due to the much higher conductivity of the lumen for water-vapor diffusion. This bending of flux toward the lumen results in a more uniform distribution than that in heat transport, such that no correction factor (Z) is necessary.

Values of D_T computed from Eqs. (6.21) and (6.22) are in reasonable agreement with those calculated by Comstock (1963), Choong (1965), Stamm and Nelson (1961), and Rosen (1978), however Eqs. (6.21) and (6.22) tend to give low values when applied to wood drying for reasons discussed previously. Choong and Skaar (1972) found significantly higher values than predicted by Eq. (6.22) when the surface resistance of thin specimens were accounted for. This is discussed in detail in Chap. 7.

6.6 The Importance of Pit Pairs in Water-Vapor Diffusion

The conductivity model described in the previous section neglects the effect of pit openings. Choong (1965) has shown by model calculations that the pit openings become increasingly more important at low moisture contents and high specific gravities where the higher values of D_v and greater cell-wall resistance make the effect of pit openings more significant. Since the drying of wood is usually done at higher moisture contents, and since pit aspiration occurs before the wood-moisture content goes below the fiber-saturation point, there is some justification for neglecting the contribution of the pit openings.

Some simple calculations may be made to estimate the importance of the pit-opening path under extreme conditions. Petty (1973) investigated the diffusion of nonswelling gases through *Picea sitchensis* and *Abies grandis* and found that, in the tangential direction, 90% of the resistance to diffusion was in the pit apertures, 10% in the membrane pores, and 0.3% in the lumens. Therefore, only the relatively high resistance of the apertures will be considered. It is assumed that the chambers have 1/3 the diameter of the lumen, the torus is 1/2 to 1/3 the diameter of the chamber, and the aperture is 1/2 the diameter of the torus. Therefore, the aperture is 1/12 to 1/18 the diameter of the lumen, with an average of 1/15. Assuming $L/d = 100$ for the cell, 50 pit pairs per tracheid, and that the length of the aperture path is equal to the double cell-wall thickness. The ratio of the aperture area to the lumen area is then $50 \, \pi(d/15)^2 / 400 \, \pi \, d^2 = 0.0006 = 0.06\%$. At a moisture content of 5%, $D_V/D_{BT} \approx 2,000$ from Fig. 6.5. Tarkow and Stamm (1960) estimated the coefficient of hindered diffusion in the pits to be 1/40 of that in the lumens. Therefore the contribution of the pits to water vapor diffusion through the cell wall on a basis of relative area, assuming that all pits and tracheids are conducting, is 0.06 (2,000)(1/40) = 3.0%. Petty (1973) estimated that the ratio for hindered diffusion is 1/3 rather than 1/40. Then, according to Petty's estimate, the contribution of the pit openings would be 0.06 (2,000)(1/3) = 40%. Although the latter estimate is very significant, it is based on full conduction through all pit pairs and tracheids. This could only occur in the most permeable softwoods. Petty (1970) found 7% of the tracheids conducting in *Picea sitchensis* sapwood which would significantly reduce the estimate of 40%. He estimated approximately 5 to 6 conducting pit pairs per tracheid in agreement with the findings of Phillips (1963). Usually aspiration closes most of the earlywood pits and this will have occurred before the moisture content is reduced below the fiber saturation point. When these factors are considered, it is probably reasonable to neglect the contribution of pit openings to transverse water-vapor diffusion, even at low moisture contents.

It is not expected that pit-opening conductivity would be an important factor in the longitudinal direction because most of the resistance to diffusion is in the lumens. Assuming the length of the lumen path as 100 d and that of the pit-aperture path as d/4, the area of the lumen as πd^2, and 1/3 for hindered diffusion, the relative resistance of the apertures $= \dfrac{d/4}{100 \, d} \cdot \dfrac{\pi \, d^2 \, (3)}{50 \, \pi \, (d/15)^2} = 0.034 = 3\%$. Therefore most of the resistance is in the lumens and the relatively high conductance of the pit pairs will have a negligible effect. Petty (1973) found 1% of the resistance in the apertures for longitudinal diffusion, with 98% in the lumens.

At temperatures above the boiling point and at moisture contents above the fiber saturation point, hydrodynamic bulk flow results from the pressure gradient arising from steam pressure within the wood. Under these conditions the permeability of the wood may be the dominant factor controlling the moisture movement. The Poiseuille flow combined with increased bound-water diffusion results in a sharp increase in the rate of flow, which cannot be readily calculated due to the complexity of the process. In addition, the effect of capillary forces may be significant above the fiber saturation point (Chap. 4.8).

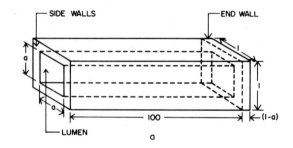
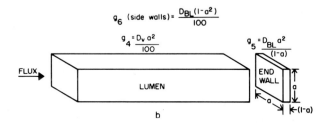

Fig. 6.7. Longitudinal model for moisture movement in wood. **a** Cellular geometrical model for longitudinal flow; **b** significant elements of geometrical model

6.7 Longitudinal Moisture Diffusion Model

Due to the relatively low diffusion coefficient of cell-wall substance (D_{BL}) compared with D_V, the end walls of the cells cannot be neglected as they were in the derivation of the equation for longitudinal thermal conductivity. A square cross section is assumed with a length to diameter ratio of 100. The longitudinal model is illustrated in Fig. 6.7.

The lumens and side walls form a parallel circuit which is in series with the end walls. Therefore,

$$\frac{1}{g_L} = \frac{1}{g_4 + g_6} + \frac{1}{g_5}, \tag{6.23}$$

where g_L = longitudinal water-vapor conductance of wood.

Due to the very high ratio of D_V/D_{BL}, the conductance of the side walls is negligible in comparison with that of the lumens, especially at low moisture contents. Therefore the term g_6 may be eliminated from Eq. (6.23) and it may be simplified to

$$\frac{1}{g_L} = \frac{1}{g_4} + \frac{1}{g_5}. \tag{6.24}$$

In order to express the conductance, g_L, in terms of D_L, the factor $(1-v_a)$ or $(1-a^2)$ must be multiplied by D_L to convert the concentration gradient from the gross wood to the cell-wall basis. In addition D_L must be divided by 100 because the length is 100 units. Then

$$g_L = D_L (1-a^2)/100.$$

Similarly, $g_4 = D_V a^2/100$, and $g_5 = D_{BL} a^2/(1-a)$.

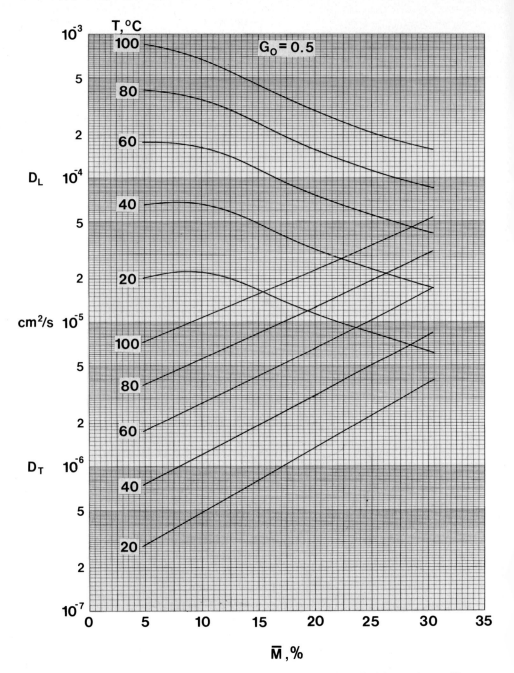

Fig. 6.8. Longitudinal and transverse moisture diffusion coefficients of wood with ovendry specific gravity of 0.5, calculated from Eqs. (6.21), (6.22), (6.25), and (6.26)

Then, substituting into Eq. (6.24) and solving for D_L,

$$D_L = \frac{a^2}{(1-a^2)} \frac{D_v D_{BL}}{D_{BL} + 0.01(1-a)D_v}. \tag{6.25}$$

$M < 20\%$

Since the conductance of the side walls was neglected in the derivation of Eq. (6.25), this equation results in low values of D_L, particularly at high moisture contents (above 20%) where the conductance of the side walls becomes relatively more significant as the value of D_{BL} increases and D_v decreases as revealed in Fig. 6.5. When all three conductances (g_4, g_5, and g_6) are substituted into Eq. (6.23), a more complex equation is obtained which is valid at all moisture contents up to the fiber saturation point. For greater simplicity it is written in terms of the reciprocal diffusivity.

$$\frac{1}{D_L} = (1-a^2) \left[\frac{1}{D_v a^2 + D_{BL}(1-a^2)} + \frac{(1-a)}{100 D_{BL} a^2} \right]. \tag{6.26}$$

$M < M_f$

Figure 6.8 is a plot of D_T and D_L at various temperatures and moisture contents for a wood with an oven-dry specific gravity of 0.5. The ratio D_L/D_T varies from approximately 100 at 5% moisture content to 2 to 4 at 30%. It is clear when this figure is compared with Fig. 6.5 that the values of D_T are approximately directly proportional to D_{BT} and that the diffusion coefficient of the air in the lumens, D_v, is the dominant influence resulting in the decrease of D_L with increasing moisture content.

6.8 Nonisothermal Moisture Movement

There has been very little reference in the wood-science literature to thermal diffusion of water vapor in wood. Some insights into its importance are provided by the nonisothermal experiments of Babbitt (1940), Voight et al. (1940), and Choong (1963). Babbitt investigated fiber board, and his results may not be typical of wood in the transverse direction due to continuous openings in which gaseous interdiffusion would dominate. He found a significant flux with a temperature difference of only 10 °C across the specimen and with small gradients of relative humidity and moisture content. Thermal gaseous interdiffusion is discussed at length by de Groot (1962), Katchalsky and Curran (1965), and Jost (1960). It is not strongly affected by a thermal gradient due to the relatively weak dependence of the coefficient on temperature (approximately proportional to $T^{1.5}$).

The experiments of Voight et al. (1940) and Choong (1962) were discussed by Siau (1980). These nonisothermal experiments were performed with relatively steep thermal gradients of approximately 10 °C/cm and the specimens were encapsulated to prevent net moisture movement to the outside and to assure a net flux of zero within the specimens at equilibrium. After equilibrium had been established, the specimens were sliced to determine the moisture content and partial-vapor-pressure profiles between two parallel surfaces across which a temperature gradient was applied. In all cases the results indicated the highest moisture content on the cool side and the highest partial vapor pressure on the warm side. The application of Fick's first law would predict a net flux from the cool to the warm side using a gradient of moisture content and a flux in the reverse direction based upon a gradient of partial vapor pressure. Clearly neither of these explains the observed phenomena.

Nonisothermal moisture movement can be analyzed by two methods: (a) as due to a gradient of activated moisture molecules, and (b) as due to a gradient of chemical potential (or water potential). Equations will be derived for each approach and comparisons made.

Skaar and Siau (1981) proposed an equation based upon a gradient of activated moisture molecules, which is applicable only for transverse diffusion. Fick's first law may be modified to the form

$$J = -K_{M*} (dM*/dx), \tag{6.27}$$

where K_{M*} = coefficient for diffusion of activated molecules = $K_M(\partial M/\partial M*)$, and $M*$ is the content of activated moisture molecules in the wood based on oven-dry weight. This can be calculated from the Boltzmann distribution as

$$M* = M \exp(-E_b/RT), \tag{6.28}$$

where E_b = activation energy, defined by Eq. (6.13). Since $M*$ is a function of both M and T, the gradient may be evaluated as:

$$dM*/dx = (\partial M*/\partial T)(dT/dx) + (\partial M*/\partial M)(dM/dx). \tag{6.29}$$

Evaluating the partial derivatives, assuming that E_b is independent of T,

$$\frac{\partial M*}{\partial T} = \frac{ME_b}{RT} \exp(-E_b/RT), \tag{6.30}$$

$$\frac{\partial M*}{\partial M} = \left(1 - \frac{M}{RT}\frac{\partial E_b}{\partial M}\right) \exp(-E_b/RT). \tag{6.31}$$

Since $\partial E_b/\partial M = -70$ cal/(mol %) from Eq. (6.13), Eq. (6.31) may be rewritten as

$$\frac{\partial M*}{\partial M} = \frac{RT + 70 M}{RT} \exp(-E_b/RT). \tag{6.32}$$

By substitution, the equation based upon a gradient of activated molecules may be written as

$$J = -K_M\left[\left(\frac{M}{RT+70 M}\right)\left(\frac{9,200-70 M}{T}\right)\frac{dT}{dx} + \frac{dM}{dx}\right], \tag{6.33}$$

where $K_M = K_{M*}(\partial M*/\partial M)$. The coefficient K_M is then the same as that used in the isothermal case in Eqs. (6.4b) and (6.5).

The coefficient K_M may be calculated from D_T by combining Eqs. (6.5), (6.14), and (6.21) because Eq. (6.33) is only applicable to transverse water vapor movement. The coefficient, K_M, may then be calculated as

$$K_M = \frac{0.0007 \, G\varrho_w \exp[-(9,200-70 M)/RT]}{(1-a^2)(1-a)}. \tag{6.34}$$

Equation (6.33) was used by Siau and Babiak (1983) to analyze the results of a series of steady-state nonisothermal experiments in which the moisture flux through a 4.74-cm-thick specimen of *Pinus strobus* was monitored by successive weighings of a diffusion cup on the cool side of the specimen. A thermal gradient of 7.2 °C/cm was used. The experimental results were in reasonable agreement with the theoretical values calculated from Eq. (6.33). The results could not be explained by using the isothermal forms of Fick's first law for either a gradient of moisture content or partial vapor pressure.

An alternative nonisothermal equation has been proposed by Siau (1983a) based on a gradient of the temperature-dependent chemical or water potential. Starting with Fick's first law for a chemical-potential gradient,

$$J = -K_\mu \frac{d\mu}{dx}, \tag{6.35}$$

where

$$K_\mu = K_M \left(\frac{\partial M}{\partial \mu}\right)_T = K_M \left(\frac{\partial H}{\partial \mu}\right)_T \left(\frac{\partial M}{\partial H}\right)_T. \tag{6.36}$$

and μ represents the chemical potential of the water vapor in unsaturated air and in wood below fiber saturation point in equilibrium with each other as described by Eq. (4.11).

Since chemical potential changes with both T and M (or relative humidity), Eq. (6.35) may be rewritten in the form,

$$J = -K_\mu \left(\frac{\partial \mu}{\partial T} \frac{dT}{dx} + \frac{\partial \mu}{\partial M} \frac{dM}{dx}\right). \tag{6.37}$$

Substituting H/100 for h in Eq. (4.11) and evaluating the derivatives,

$$\frac{\partial \mu}{\partial T} = \frac{\partial \mu_1^0}{\partial T} + RT\left(\frac{\partial \ln H/100}{\partial T}\right) + R \ln H/100. \tag{6.38}$$

Recalling Eq. (4.12),

$$\frac{\partial \mu_1^0}{\partial T} = 10.37 + 0.0077\,(T - 273), \tag{6.39}$$

also,

$$\frac{\partial \mu}{\partial M} = \left(\frac{\partial \mu}{\partial H}\right)_T \left(\frac{\partial H}{\partial M}\right)_T, \tag{6.40}$$

where

$$\left(\frac{\partial \mu}{\partial H}\right)_T = \frac{RT}{H}; \tag{6.41}$$

then, by substitution,

$$J = -K_M \left[\frac{H}{RT}\left(\frac{\partial M}{\partial H}\right)_T \left(\frac{\partial \mu_1^0}{\partial T} + RT\frac{\partial \ln H/100}{\partial T} + R \ln H/100\right)\frac{dT}{dx} + \frac{dM}{dx}\right]. \tag{6.42}$$

The term $\partial \ln (H/100)/\partial T$ may be evaluated from sorption isotherm data for the wood at a series of temperatures. If these data are not available, an assumption may be based upon the general data available in the Wood Handbook (USDA 1955). The above term is related to the molar differential heat of sorption of wood, E_L, defined by Skaar (1972) as

$$E_L = -R\left(\frac{\partial \ln H/100}{\partial (1/T)}\right)_M = RT^2\left(\frac{\partial \ln H/100}{\partial T}\right)_M. \tag{6.43}$$

The differential heat of sorption is the difference between the molar heat of vaporization of bound water in wood (E_v) and the molar heat of vaporization of free water (E_0) at the same temperature. It may be determined from the heat of wetting as described by Skaar (1972).

When E_L is calculated from the Wood Handbook (USDA 1955) data using Eq. (6.43), the variation with moisture content approximately follows an empirical equation, the form of which was suggested by Skaar (1972)

$$E_L = 5{,}000 \exp(-0.14\,M). \tag{6.44}$$

In order to facilitate the use of Eq. (6.42) to characterize nonisothermal moisture movement, it is helpful to fit the sorption isotherm data to the Hailwood-Horrobin equations in the manner used by Simpson (1973). Simpson has fitted the Wood Handbook data to these equations, making it possible to evaluate $(\partial M/\partial H)_T$ and E_L. If E_L is substituted, Eq. (6.42) may be simplified to

$$J = -K_M\left[\frac{H}{RT}\left(\frac{\partial M}{\partial H}\right)\left(\frac{\partial \mu_1^0}{\partial T} + \frac{E_L}{T} + R\ln H/100\right)\frac{dT}{dx} + \frac{dM}{dx}\right]. \tag{6.45}$$

The Hailwood-Horrobin equation for M(H) has the form:

$$M = \frac{1{,}800}{W}\left(\frac{K_1 K_2 H}{100 + K_1 K_2 H} + \frac{K_2 H}{100 - K_2 H}\right). \tag{6.46}$$

The parameters are functions of T (Simpson 1973) and may be evaluated as

$$W = 216.9 + 0.01961\,F + 0.00572\,F^2, \tag{6.47}$$

$$K_1 = 3.73 + 0.03642\,F - 0.0001547\,F^2, \tag{6.48}$$

$$K_2 = 0.674 + 0.001053\,F - 1.74 \times 10^{-6}\,F^2, \tag{6.49}$$

where F = Fahrenheit temperature = 9/5 (T − 273) + 32.

In some cases it is more convenient to use the form:

$$\frac{H}{M} = A + BH - CH^2, \tag{6.50}$$

where the parameters may be evaluated as:

$$A = \frac{W}{18}\left[\frac{1}{K_2(K_1+1)}\right], \tag{6.51}$$

$$B = \frac{W}{1,800}\left(\frac{K_1-1}{K_1+1}\right), \tag{6.52}$$

$$C = \frac{WK_1K_2}{180,000(K_1+1)}. \tag{6.53}$$

Then, solving Eq. (6.50) for M,

$$M = \frac{H}{A+BH-CH^2} \tag{6.54}$$

and, for H, using the solution for a quadratic equation,

$$H = \frac{MB-1+\sqrt{(1-MB)^2+4CM^2A}}{2CM}; \tag{6.55}$$

Also, from Eq. (6.54),

$$\left(\frac{\partial M}{\partial H}\right)_T = \frac{A+CH^2}{(A+BH-CH^2)^2}. \tag{6.56}$$

In summary, it may be concluded that the Hailwood-Horrobin equation has been found to produce a good fit to wood sorption-isotherm data. The use of these equations makes it possible to evaluate H from M and the derivative $(\partial M/\partial H)_T$ which are necessary in the application of Eq. (6.45) to nonisothermal experimental data.

It is clear that Eqs. (6.33) and (6.45) have the same form. The coefficients, K_M, are equal and dM/dx appears in both. Since they describe the same phenomena, the multipliers of dT/dx, designated as $(dM/dT)_{M*}$ and $(dM/dT)_\mu$, respectively, should be equal. The subscripts are used to indicate a gradient of activated molecules (M*) or of chemical potential (μ).

The difference between results obtained for nonisothermal calculations based upon Eqs. (6.33) and (6.45) may be examined by evaluation of $(dT/dM)_{M*}$ and $(dT/dM)_\mu$ at various temperatures and moisture contents. This has been done by Siau (1983a) and $(dT/dM)_\mu$ is higher at low relative humidities and moisture contents while the reverse is true at high relative humidities. The two functions become equal at H=53%. This trend is similar at temperatures between 0 °C and 100 °C. A plot of the results at 40 °C is revealed in Fig. 6.9, since this is a typical result. This indicates that a temperature difference of approximately 17 °C is equivalent in its effect on flux to a 1% difference in moisture content at H=10%, while the value is only 1.2 °C at H=90%, when chemical potential is used as the gradient, while the corresponding values for a gradient of activated molecules are 11 °C at H=10% and 4 °C at H=90%. Thus in both cases, the effect of the thermal gradient upon the flux becomes more important as the relative humidity or EMC of the wood increases. These facts are clarified in Fig. 6.10, in which the ratios of $(dM/dT)_\mu$

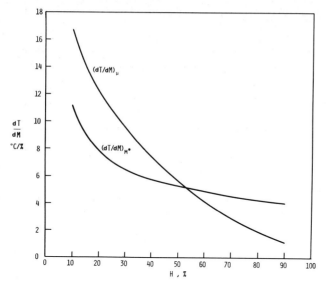

Fig. 6.9. A comparison of dT/dM based upon a gradient of chemical potential to that based upon a gradient of activated molecules as a function of the relative humidity of the air in equilibrium with the wood surface at 40 °C. (Siau 1983a)

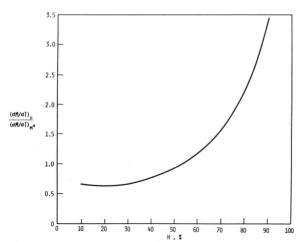

Fig. 6.10. The ratio of dM/dT based upon a gradient of chemical potential to that based upon a gradient of activated molecules as a function of the relative humidity of the air in equilibrium with the wood surface at 40 °C. (Siau 1983a)

to $(dM/dt)_{M*}$ are plotted. In this case the ratio extends from 0.65 at H=10% to 1.0 at H=53% and to 3.4 at H=90%. It is evident from this that the use of the chemical potential rather than the activated moisture equation accentuates the influence of the thermal term at high wood-moisture contents and decreases it at low moisture contents. At the point of equivalence (H=53%), a temperature differential of approximately 5 °C is equivalent in its effect to a 1% moisture-content difference.

Nonisothermal moisture movement is important in heated buildings with thick wood walls such as log homes. Under such conditions, the exterior walls are at a

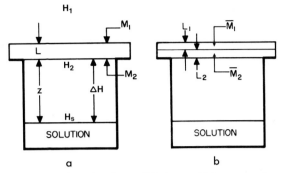

Fig. 6.11. Illustrating two methods for the determination of moisture diffusion coefficients using the diffusion cup or vapometer. **a** Method 1 using one specimen. Gradient $= (M_2-M_1)/L$, where M_1, $M_2 =$ moisture contents in equilibrium with the relative humidities at the two surfaces, H_1 and H_2. $\bar{M} = (M_1+M_2)/2$. **b** Method 2 using two matched specimens. Gradient $= (\bar{M}_2-\bar{M}_1)/\bar{L}$, where $\bar{M}_1$, $\bar{M}_2 =$ moisture contents of the upper and lower veneers, $\bar{L} =$ average thickness, $\bar{M} = (\bar{M}_1 + \bar{M}_2)/2$

low temperature and high relative humidity, giving an EMC of approximately 12% to 15%. The interior, however, may be at a low EMC of approximately 6% to 8% due to the low relative humidity in the heated interior. Under these conditions, Eq. (6.33) predicts a net flux of moisture from the outside to the inside due to the overriding effect of the moisture-content difference. Also, nonisothermal moisture movement is important in wood drying since a thermal gradient is required in this process.

6.9 Measurement of Diffusion Coefficients by Steady-State Method

Method 1, Using One Specimen

A vapometer or diffusion cup is used as depicted in Fig. 6.11 a. Water for 100% relative humidity or a saturated salt solution for lower values is used in the cup, preferably agitated by a magnetic stirrer to maintain the desired relative humidity. The specimen is edge-sealed with paraffin or epoxy. The assembled cup is placed in a controlled-humidity chamber or room. As moisture migrates through the specimen, the flux may be monitored by weighing the complete cup. Steady-state conditions are indicated by a linear relationship between weight change and time. The flux is then equal to the rate of weight change (g/s) divided by area of the specimen exposed to moisture movement.

When one specimen is tested in the vapometer, it is necessary to utilize the sorption isotherm in the determination of the moisture-content gradient. If this is not available, the Wood Handbook (USDA 1955) data may be used as an approximation. The relative humidities on each surface of the specimen must be known. It is assumed that equilibrium exists on both surfaces, therefore the EMC's may be taken as the moisture contents (M_1 and M_2). The relative humidity of the conditioning chamber (H_1) is controlled and the relative humidity directly over the surface of the solution (H_s) may be determined from Table 1.6 or 1.7. Due to the flux in the cup, there is a differential of relative humidity (ΔH) which must be known to find the relative humidity on the lower surface of the specimen (H_2). In specimens of very low water-vapor conductance, this gradient may be negligible and H_2 may be taken as equal to that in equilibrium with the solution. However, an evaluation of ΔH must usually be made to determine whether it is negligible.

The value of ΔH may be calculated from the moisture-concentration gradient in the cup by use of the general gas law. The concentration gradient may be determined from the flux and coefficient as

$$\frac{\Delta(w/V)}{z} = \frac{\text{flux}}{D_a}, \qquad (6.57)$$

where $\Delta(w/V)$ = difference in water vapor concentration (or absolute humidity) between the surface of the solution and the underside of the wood specimen, g/cm³, and z = distance from surface of solution to underside of specimen, cm.

From the general gas law, it may be shown that

$$\Delta(w/V) = \frac{18\Delta p}{RT}, \qquad (6.58)$$

where Δp = partial vapor pressure potential between solution and wood specimen, cm Hg, R = universal gas constant = 6,230 cm³ cm Hg/mol K.

By substituting Eq. (6.57) into (6.58),

$$\Delta p = \frac{z(\text{Flux})}{D_a}\left(\frac{RT}{18}\right), \qquad (6.59)$$

The change in relative humidity corresponding to Δp is then

$$\Delta H = \frac{100\, z(\text{Flux})}{p_0 D_a}\left(\frac{RT}{18}\right). \qquad (6.60)$$

From Fig. 6.11a, $H_2 = H_s - \Delta H$, D_a is calculated from Eq. (6.15), and the value of M_2 on the underside is then determined from H_2 using the sorption isotherm. Similarly, M_1 may be determined from the relative humidity of the chamber, H_1, or it may be determined from a matched specimen which is equilibrated in the controlled chamber. The cup is then disassembled and the moisture content, specific gravity, and thickness of the specimen (L) are measured, and the gradient in the specimen calculated as $(M_2-M_1)/L$. The value of K_M or D corresponding to the average moisture content $(M_2+M_1)/2$ is calculated from Eq. (6.4b) or (6.5).

Method 2, Using Two Matched Specimens

The procedure is similar to Method 1, except that two matched specimens are used, as shown in Fig. 6.11b. After equilibrium has been established, the cup is disassembled and the moisture contents of the two specimens ($\bar{M}_1$ and $\bar{M}_2$) are determined by drying. The gradient is then taken as $(\bar{M}_2-\bar{M}_1)/\bar{L}$, where $\bar{L}$ = average thickness. The average moisture content for the coefficient is then $(\bar{M}_2+\bar{M}_1)/2$ and the coefficients are calculated as in Method 1. The sorption isotherm and ΔH are not required in this method because the moisture contents are measured directly. Alternatively, the coefficient may be determined from the sorption isotherm and ΔH, using equilibrium values of M_1 and M_2 for the outer surfaces as in Method 1.

Table 6.1. Permeance of construction materials

Material	Permeance, Perms
15-lb tar felt	4.0
Aluminium foil, 1 mil	0.0
Polyethylene (2 mil)	0.16
Polyethylene (6 mil)	0.06
Still air, 1 inch thick	120.
Concrete, 1 inch thick	3.2
Concrete block, 8 inch	2.4
Hardboard, 1/8 inch standard	11.0
Wood, 1 inch thick	0.12
Wood, 8 inches thick	1.0

Method 3, ASTM Method for Vapor Barriers

The movement of water vapor through materials used as vapor barriers in construction may be evaluated using the procedures described in specifications issued by ASTM (1968) both for materials less than one-eight inch thick and for thicker materials. Diffusion cups are used as illustrated in Fig. 6.11 a. Various conditions are specified but usually a temperature of 23 °C, with the relative humidity of the controlled chamber maintained at 50%, with that in the cup 100% when distilled water is used, or 0% when a desiccant such as anhydrous calcium chloride is utilized. Some tests are performed with the side in the cup wetted by inverting the cup. Air velocity, in meters per minute, in the conditioned chamber is specified to be not less than five times the permeance in metric perms and shall not exceed 150 m/min. A coefficient called permeance is determined which is defined as

$$\text{Permeance} = w/(tA\Delta p), \tag{6.61}$$

where w/t = rate of water-vapor transport, g/day or grains/hour, A = area of opening in cup, m^2 or ft^2, Δp = vapor pressure differential, mm Hg or in Hg.

When metric units are used in the calculations, permeance is expressed in metric perms, while that in English units is in perms. It can be shown that

1 metric perm = 1.52 perm or 1 perm = 0.66 metric perm.

A material is considered a good vapor barrier if its permeance is less than one perm or one metric perm. Some typical values for construction materials are provided in Table 6.1.

Permeance can be related to the water-vapor diffusion coefficient based upon a partial vapor pressure gradient (K_p) of a hygroscopic material such as wood by taking its thickness into account to determine its water-vapor permeability.

$$\text{Water-vapor permeability} = \frac{wL}{tA\Delta p} \tag{6.62}$$

Water-vapor permeability may be expressed in metric perm cm. When cgs units are used in Eq. (6.62), the result is K_p, the coefficient for water-vapor transport based upon a partial vapor-pressure gradient, expressed in gm/(cm s cm Hg).

It is clear then that

$$\frac{K_p}{L} = \text{metric perm} \times \frac{1 \text{ day}}{86{,}400 \text{ s}} \times \frac{1 \text{ m}^2}{10^4 \text{ cm}^2} \times \frac{10 \text{ mm Hg}}{1 \text{ cm Hg}}$$

$$= \text{metric perm} \times 1.16 \times 10^{-8}$$

Then

$$K_p = \text{metric perm} \times L \times 1.16 \times 10^{-8}, \quad (6.63)$$

where L is expressed in cm.

Also, the coefficient D may be calculated from Eq. (6.6) as

$$D = \frac{K_p p_0}{G \varrho_w} \frac{\partial H}{\partial M}, \quad (6.64)$$

where p_0 is expressed in cm Hg.

Then, by substitution,

$$D = \frac{\text{metric perm} \times 1.16 \times 10^{-8} L p_0}{G \varrho_w} \frac{\partial H}{\partial M}, \quad (6.65)$$

where L is expressed in cm, p_0 in cm Hg.

It is evident from Table 6.1 that wood becomes a suitable vapor barrier at a thickness of approximately 8 inches. However, there is no need for a vapor barrier inside a thick wood exterior wall because the interior dew point temperature is located within the wood and therefore there is no cold surface on which the moist interior air can condense. Roofing felt is inadequate as a vapor barrier but is suitable for preventing the flow of liquid water through a roof. Polyethylene sheet and metal foil are both very effective as vapor barriers provided they have not been punctured during installation.

All of the methods of steady-state diffusion coefficient determination discussed above are based upon Fick's law in which a moisture-content or partial pressure gradient may be used with equivalent results provided conditions are isothermal. This is probably the case with thin vapor barriers, but would not be applicable to thick barriers where activated diffusion takes place such as through wood walls in the transverse direction. In addition to this, no consideration has been given to surface losses which have the greatest significance in thin specimens. Skaar (1954), Choong and Skaar (1972) have shown that surface resistance can cause the results to be low as much as 40% for a thickness of 3 cm. Rosen (1978) has shown, in addition, that air velocity plays an important role with a rapid increase in adsorption rate up to 3 m/s. This increase is attributed to a decrease in the surface resistance. This subject of surface emission losses will be discussed in detail in Chap. 7.3.1 where the unsteady-state method for measuring diffusion coefficients will be described. With this method it is possible to evaluate the effect of surface resistance from measurements of matched specimens of different thicknesses.

Chapter 7
Unsteady-State Transport

7.1 Derivation of Unsteady-State Equations for Heat and Moisture Flow

Unsteady-state flow occurs when the flux and gradient are variable in both space and time. Such flow has more importance in wood treatments than steady-state flow, since it is present whenever a wood specimen is heated, evacuated, impregnated with liquids, or dried. In all these cases there is a net change in the conditions inside the specimen over a period of time.

The unsteady-state equations may be derived from the steady-state relationships. The steady-state equation for heat flow (Fourier's law) may be expressed in differential form to be applicable to the flow of heat through the thin slab depicted in Fig. 7.1.

$$\frac{\Delta H}{\Delta t} = -K_g A \frac{\Delta T}{\Delta x}, \tag{7.1}$$

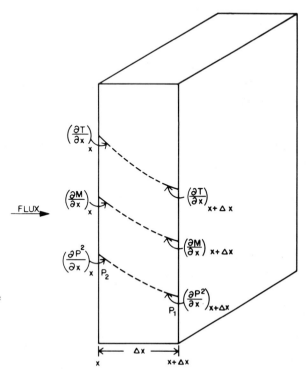

Fig. 7.1. Thin slab illustrating the unsteady-state transport of heat, moisture, and gases. *Curved lines* are profiles of temperature, moisture content, and pressure-squared

where $\Delta H/\Delta t$ = rate of heat flow, cal/s, $\Delta T/\Delta x$ = temperature gradient, °C/cm, Δx = thickness of slab, cm, A = area of slab, cm², K_g = thermal conductivity of gross wood, cal/(cm °C s).

The symbol x is used to denote variable distance in the direction of flow as distinguished from L, which represents the total length in the direction of flow in Eq. (5.2). The negative sign is used in Eq. (7.1) because heat flows in the direction of decreasing temperature.

Since ΔH and ΔT are variable in both space and time in unsteady-state flow, Eq. (7.1) may be rewritten in partial derivative form as it applies to the flow of heat into and out of the slab depicted in Fig. 7.1.

$$\left(\frac{\partial H}{\partial t}\right)_{in} = -K_g A \left(\frac{\partial T}{\partial x}\right)_x$$

$$\left(\frac{\partial H}{\partial t}\right)_{out} = -K_g A \left(\frac{\partial T}{\partial x}\right)_{x+\Delta x}.$$

The rate of heat gain into the slab, assuming K_g is constant with temperature, is then calculated as

$$\left(\frac{\partial H}{\partial t}\right)_{gain} = K_g A \left[\left(\frac{\partial T}{\partial x}\right)_x - \left(\frac{\partial T}{\partial x}\right)_{x+\Delta x}\right].$$

Thus it is the change in gradient across the thickness of the slab which determines the rate at which heat is gained. This change may be evaluated mathematically by differentiating the gradient relative to x and multiplying the resulting second derivative by the thickness Δx.

$$\left(\frac{\partial H}{\partial t}\right)_{gain} = K_g A \left[\frac{\partial(\partial T/\partial x)}{\partial x}\right]\Delta x = K_g A \left(\frac{\partial^2 T}{\partial x^2}\right)\Delta x. \tag{7.2}$$

The rate of heat gain may also be calculated as the product of the mass, specific heat, and rate of change of the average temperature of the slab.

$$\left(\frac{\partial H}{\partial t}\right)_{gain} = c\varrho A \Delta x \frac{\partial T}{\partial t}, \tag{7.3}$$

where c = specific heat of wood, cal/g °C, ϱ = density of wood, g/cm³.

Equating (7.2) and (7.3),

$$\frac{\partial T}{\partial t} = \frac{K_g}{c\varrho} \frac{\partial^2 T}{\partial x^2}. \tag{7.4}$$

Then

$$D_h = \frac{K_g}{c\varrho}, \tag{7.5}$$

where D_h = thermal diffusivity, cm²/s, $c\varrho$ = quantity of heat required to increase the temperature of a unit cube of wood by 1 °C.

Equation (7.4) may then be written

$$\frac{\partial T}{\partial t} = D_h \frac{\partial^2 T}{\partial x^2}. \tag{7.6}$$

The specific heat of moist wood may be calculated by treating wood as an additive mixture of dry cell-wall substance and water. Skaar (1972) states that the specific heat of dry wood increases significantly with temperature similar to other organic materials. He tabulated the results of six investigations in which an average value for dry wood was $(0.268 + 0.0011\text{C})$ cal/g °C, where C=Celsius temperature. Skaar also discusses a small increase in the specific heat of bound water with moisture content between 5% and FSP, but, for simplicity, the value of 1.0 cal/g °C is assumed for both bound and free water. Then the specific heat of moist wood may be calculated as

$$c = \frac{0.268 + 0.0011\,C + 0.01\,M}{1 + 0.01\,M}. \tag{7.7}$$

The thermal diffusion coefficient may then be written as

$$D_h = \frac{K_g(1 + 0.01\,M)}{\varrho(0.268 + 0.0011\,C + 0.01\,M)} \tag{7.8}$$

or, in terms of specific gravity, by substitution of Eq. (1.17) into (7.8):

$$D_h = \frac{K_g}{G(0.268 + 0.0011\,C + 0.01\,M)\varrho_w}, \tag{7.9}$$

where K_g is calculated using one of the equations presented in Chap. 5, ϱ_w = normal density of water = 1 g/cm³.

When Eq. (5.5) is substituted into (7.9), the following result is obtained:

$$D_{h_T} = \frac{G(4.80 + 0.090\,M) + 0.57}{G(0.268 + 0.0011\,C + 0.01\,M)\varrho_w} \times 10^{-4}, \tag{7.10}$$

$M < 40\%$

where D_{h_T} = transverse thermal diffusion coefficient, cm²/s; for M greater than 40%, substitute 0.125 M for 0.090 M.

Alternatively, the empirical Eq. (5.19) may be used:

$$D_{h_T} = \frac{12.2 - 11.3\sqrt{1 - G(0.667 + 0.01\,M)}}{G(0.268 + 0.0011\,C + 0.01\,M)\varrho_w} \times 10^{-4}. \tag{7.11}$$

$M < M_f$

Since, according to Eq. (5.7), the longitudinal thermal conductivity of wood is 2.5 times the transverse value, it may be stated that

$$D_{h_L} = 2.5\,D_{h_T}, \tag{7.12}$$

where D_{h_L} = longitudinal thermal diffusion coefficient, cm²/s.

Equation (7.6) states that the rate of temperature change (time derivative of temperature) at a point within a specimen is directly proportional to the space derivative of the gradient. The diffusion coefficient may then be defined as the ratio of the time derivative of the potential (temperature, in this case) to the space derivative of the potential gradient. The thermal diffusion coefficient is numerically equal to the rate of temperature change when the gradient changes one degree per centimeter per centimeter of length in the flow direction.

A similar derivation may be performed starting with Fick's first law of diffusion. The moisture-diffusion analog of Eq. (7.2) is

$$\left(\frac{\partial w}{\partial t}\right)_{gain} = \bar{K}_M A \left(\frac{\partial^2 M}{\partial x^2}\right) \Delta x, \qquad (7.13)$$

where $\partial w/\partial t$ = rate of moisture movement, g/s, w = mass of moisture being transferred, g, M = moisture content, percent $\bar{K}_M$ = average conductivity coefficient for moisture diffusion corresponding to the average moisture content calculated as described in Chap. 6.2, g/(cm % s).

The rate of moisture gain into the slab may also be calculated as the product of the specific gravity of the wood, the normal density of water (for dimensional units), the volume, and the rate of change of moisture content (expressed as a decimal).

$$\left(\frac{\partial w}{\partial t}\right)_{gain} = \frac{G\varrho_w A \Delta x}{100} \frac{\partial M}{\partial t}, \qquad (7.14)$$

where G = specific gravity of wood at average moisture content of the slab.

Equating (7.13) and (7.14),

$$\frac{\partial M}{\partial t} = \frac{100 \bar{K}_M}{G\varrho_w} \frac{\partial^2 M}{\partial x^2}. \qquad (7.15)$$

This may also be written as

$$\frac{\partial M}{\partial t} = \bar{D} \frac{\partial^2 M}{\partial x^2}, \qquad (7.16)$$

where

$$\bar{D} = \frac{100 \bar{K}_M}{G\varrho_w}, \qquad (7.17)$$

in which $\bar{D}$ = average diffusion coefficient for moisture movement corresponding to the overall average moisture content, calculated as described in Chap. 6.2, cm²/s, $G\varrho_w/100$ = mass of moisture to raise the moisture content of a unit cube of wood by 1%.

Equation (7.16) is an approximation because it will be recalled that the coefficient increases significantly with moisture content. It is given here to show its similarity to Eq. (7.6) for unsteady-state heat transfer and for its possible usefulness in approximate calculations. An exact form of the equation may be derived by differentiating the coefficient.

$$\frac{\partial M}{\partial t} = \frac{\partial}{\partial x}\left(D \frac{\partial M}{\partial x}\right). \qquad (7.18)$$

It will be recalled (Eqs. 6.14 and 6.22) that $D_T = 0.07 \exp(-E_b/RT)/[1-a^2)(1-a)]$. Since $E_b = 9{,}200 - 70\,M$ cal/mol, $\partial E_b/\partial M = -70$ cal/mol. Then, evaluating the derivatives, Eq. (7.18) assumes the form:

$$\frac{\partial M}{\partial t} = \frac{0.07}{(1-a^2)(1-a)} \exp(-E_b/RT)\left[\frac{\partial^2 M}{\partial x^2} - \frac{70}{RT}\left(\frac{\partial M}{\partial x}\right)^2\right]. \tag{7.19}$$

Equation (7.19) and the approximate Eq. (7.16) are known as Fick's second law for diffusion. Isothermal conditions are assumed in both cases. A relatively simple graphical or algebraic solution is available for Eq. (7.16), as will be described. Equation (7.19) may be solved by its transformation to a finite-difference equation followed by an iterative algebraic solution using a digital computer.

7.2 Derivation of Unsteady-State Equations for Gaseous Flow in Parallel-Sided Bodies

The unsteady-state equation for gaseous flow may be derived from Darcy's law for gases [Eq. (3.4)]. The pressure differential at a given time will be represented as $(P_2 - P_1)$, corresponding to ΔP in Eq. (3.4), and the average pressure in the slab as $(P_2 + P_1)/2$, which has the symbol $\bar{P}$ in Eq. (3.4). Then, by substitution of differential values into Eq. (3.4),

$$\left(\frac{\Delta V}{\Delta t}\right)_{in} = -\frac{k_g A(P_2^2 - P_1^2)}{2\Delta x P_2} = -\frac{k_g A \Delta(P^2)}{2P_2 \Delta x}$$

where P_2 = upstream pressure at distance x; this is the pressure at which the differential volume, ΔV, is measured, P_1 = downstream pressure at distance $(x + \Delta x)$, k_g = superficial gas permeability.

Since volume is pressure-dependent, the flow of gas into the slab must be calculated on a mass basis. According to the gas law,

$$P_2 V = \frac{w}{M_w} RT, \tag{7.20}$$

where w = mass of gas, g, M_w = molecular weight of gas, and

$$\Delta w = \frac{P_2 M_w \Delta V}{RT}. \tag{7.21}$$

Then

$$\left(\frac{\Delta w}{\Delta t}\right)_{in} = -\frac{k_g A M_w \Delta(P^2)}{2RT\Delta x}.$$

This may be written in partial derivative form as

$$\left(\frac{\partial w}{\partial t}\right)_{in} = -\frac{k_g A M_w}{2RT}\left(\frac{\partial P^2}{\partial x}\right)_x.$$

Similarly, it may be shown that

$$\left(\frac{\partial w}{\partial t}\right)_{out} = - \frac{k_g A M_w}{2RT}\left(\frac{\partial P^2}{\partial x}\right)_{x+\Delta x}$$

Since the negative gradient of pressure squared is greater at x than at $(x+\Delta x)$, there is a net flow of gas into the slab which may be calculated as

$$\left(\frac{\partial w}{\partial t}\right)_{gain} = \frac{k_g A M_w}{2RT}\left(\frac{\partial^2 P^2}{\partial x^2}\right)\Delta x . \tag{7.22}$$

Let us now consider the gain in mass with time due to the increase in the average pressure in the slab with time. According to Boyle's law,

$$PV = (P + \Delta P)(V - \Delta V).$$

Since $\Delta P \Delta V$ is infinitesimal and may be neglected,

$$\Delta V = \frac{V \Delta P}{P}.$$

Let P_3 = average pressure in slab at beginning of time interval Δt and P_4 = average pressure in slab at end of time interval Δt.
Then

$$\left(\frac{\Delta V}{\Delta t}\right)_{gain} = \frac{A \Delta x v_a (P_3 - P_4)}{\bar{\bar{P}} \Delta t},$$

where $A \Delta x v_a$ = volume of slab which contains gas, $\bar{\bar{P}}$ = average pressure at which the volume ΔV is measured, averaged over both space and time.

From the gas law, Δw may be determined from the average pressure over the time interval Δt.

$$\Delta w = \frac{(P_3 + P_4) M_w \Delta V}{2RT}.$$

Then

$$\left(\frac{\Delta w}{\Delta t}\right)_{gain} = \frac{A \Delta x v_a M_w (P_3^2 - P_4^2)}{2RT\bar{\bar{P}}\Delta t}. \tag{7.23}$$

Equation (7.23) may be rewritten in partial derivative form as follows:

$$\left(\frac{\partial w}{\partial t}\right)_{gain} = \frac{A \Delta x v_a M_w}{2RT\bar{\bar{P}}}\frac{\partial P^2}{\partial t}. \tag{7.24}$$

Then, from Eqs. (7.22) and (7.24),

$$\frac{\partial P^2}{\partial t} = \frac{k_g \bar{\bar{P}}}{v_a}\frac{\partial^2 P^2}{\partial x^2} \tag{7.25}$$

or

$$\frac{\partial P^2}{\partial t} = \bar{D}_P \frac{\partial^2 P^2}{\partial x^2}, \tag{7.26}$$

where

$$\bar{D}_p = \frac{k_g \bar{\bar{P}}}{v_a}, \tag{7.27}$$

in which $\bar{D}_p$ = diffusion coefficient for hydrodynamic flow based upon the average pressure, averaged over space and time, $v_a/\bar{\bar{P}}$ = volume of gas required to raise the pressure within a unit cube of wood by 1 dyne/cm^2 or 1 atm.

Resch (1967) studied the unsteady-state flow of gases in wood in which the actual changes of pressure with time within bolts of *Pseudotsuga menziesii* were measured and compared with theoretical values obtained from the diffusion differential equation for a cylinder. A finite difference equation was used to calculate the results, and a reasonably good agreement was obtained between the experimental and theoretical values.

Prak (1970) investigated the unsteady-state flow of helium into cylindrical specimens of several species of wood in the transverse and/or longitudinal directions. Theoretical pressure-time curves were calculated from the unsteady-state equations using steady-state permeability data. It was necessary to assume the presence of two components with high and low permeabilities to achieve agreement between the theoretical curves and the experimental results, suggesting that wood is not a uniformly permeable medium. A practical example is cited of the red oaks, which have a very high longitudinal permeability, but which require a long treating time to fill all the voids, due to the relative difficulty of penetrating the fiber lumens compared with the vessels. This concept of wood components of different permeabilities is discussed in greater detail below and in Sect. 7.6.1.

Equation (7.26) is an approximate unsteady-state equation for gas flow for the same reason that Eq. (7.16) is approximate for moisture movement; namely, that the coefficient D_p is pressure-dependent due to the effect of molecular slip flow, or, in the more complex case, due to the presence of two or more conductances in series according to the Petty model. Sebastian et al. (1973) derived an exact equation in which the gas permeability is described by the Klinkenberg equation as $k_g = k(1 + b/\bar{P})$. Since permeability is a function of pressure, its value must be differentiated relative to pressure. The $\bar{P}$ term is represented as the variable pressure. Equation (7.25) may then be modified to:

$$\frac{\partial P^2}{\partial t} = \frac{1}{v_a}\left[\frac{\partial}{\partial x}\left(k_g P \frac{\partial P^2}{\partial x}\right)\right]. \tag{7.28}$$

From the Klinkenberg Eq. (3.36) $k_g P = kP + kb$ where k = constant intercept permeability.

Let dimensionless time $\tau = \dfrac{4tkP_a}{v_a L^2}$, and $X = 2x/L$,

where P_a = atmospheric pressure and L = length of the specimen in the flow direction. The constant, 4, is added because complete penetration occurs after the gas has moved a distance of $L/2$, assuming flow perpendicular to two parallel surfaces. Dimensionless distance X is then represented as $2x/L$. Pressure is also made dimensionless by dividing pressure by atmospheric pressure, since Sebastian's experiments were done between vacuum and atmospheric pressure. Then $P = P/P_a$.

In addition, b is made dimensionless by introducing $\alpha = b/P_a$. Making these substitutions Eq. (7.28) assumes the form:

$$\frac{\partial P^2}{\partial \tau} = \frac{\partial}{\partial X}\left[(P+\alpha)\frac{\partial P^2}{\partial X}\right]. \tag{7.29}$$

Then, performing the differentiation,

$$\frac{\partial P^2}{\partial \tau} = (P+\alpha)\frac{\partial^2 P^2}{\partial X^2} + \frac{\partial P}{\partial X}\frac{\partial P^2}{\partial X}. \tag{7.30}$$

(Sebastian equation for parallel-sided specimens)

Sebastian et al. (1973) solved Eq. (7.30) by transformation to a finite-difference equation and by use of a digital computer. They obtained a reasonable agreement between experimental and theoretical curves when Eq. (7.30) was applied both to the evacuation from atmospheric pressure to approximately zero absolute pressure and in subsequent filling to atmospheric pressure with air. The specimens used were cylindrical with the cylindrical surface coated allowing flow only perpendicular to the parallel ends in the longitudinal direction.

The Petty model for two series conductances can be incorporated into Eq. (7.28). It will be recalled that a curvilinear relationship between k_g and $1/\bar{P}$ can be explained by two conductances in series with each of them obeying the Klinkenberg equation (Chap. 3.10.2). Permeability is then calculated as:

$$k_{gL} = \frac{(A+l/\bar{P})(B+m/\bar{P})}{A+B+l/\bar{P}+m/\bar{P}}, \tag{7.31}$$

where A and B are intercept conductances and l and m are the respective slopes of the plots of conductance vs. reciprocal pressure. The parameters A, B, l, and m may be calculated by means of a gradient-search or nonlinear curve-fitting computer program (Dye and Nicely 1971) from experimental data of k_g vs $1/\bar{P}$. When k_g, as defined by Eq. (7.31) is substituted into Eq. (7.28) with dimensionless parameters, and the differentiation performed, the following equation is obtained:

$$\frac{\partial P^2}{\partial \tau} = \frac{A+B}{AB}\frac{(m+BP)(l+AP)}{m+l+AP+BP}\frac{\partial^2 P^2}{\partial X^2}$$

$$+ \frac{(m+l+AP+BP)(mA+lB+2ABP^2)-(ml+mAP+lBP+ABP^2)(A+B)}{(m+l+AP+BP)^2}$$

$$\times \frac{\partial P}{\partial X}\frac{\partial P^2}{\partial X}. \tag{7.32}$$

(Unsteady-state gas-flow equation using Petty model where P is dimensionless.)

Siau (1976) explains the application of this equation to experimental results with curvilinear relationships between k_g and $1/\bar{P}$ and was able to achieve some good agreements between experimental and theoretical results when this was combined with the assumption of parallel zones within the wood with different permeabilities. This will be discussed further in Sect. 7.6.1.

A simplified modification of Eq. (7.25) may be obtained by referring back to Eq. (7.23) of the derivation. This contains the term $P_3^2 - P_4^2$, which may be factored into $(P_3-P_4)(P_3+P_4)$. The latter divided by two is equal to the average pressure over the time interval Δt. If this is assumed equal to $\bar{P}$, the average pressure at which ΔV is measured, then Eq. (7.25) may be written as

$$\frac{\partial P}{\partial t} = \frac{k_g}{2v_a}\frac{\partial^2 P^2}{\partial x^2}. \tag{7.33}$$

When k_g has the form of the Klinkenberg equation, is differentiated, and the variables transformed into dimensionless form, Eq. (7.33) has the form:

$$\frac{\partial P}{\partial \tau} = 0.5\left[\left(1+\frac{\alpha}{P}\right)\frac{\partial^2 P^2}{\partial X^2} - \frac{\alpha}{P^2}\frac{\partial P}{\partial X}\frac{\partial P^2}{\partial X}\right]. \tag{7.34}$$

The author has made a comparison of Eqs. (7.30) and (7.34), using the same data for each after transformation to finite-difference equations. The results were in relatively good agreement but were not identical. There is not sufficient experience with Eq. (7.34) to enable one to decide if it is preferable to Eq. (7.30).

The equations derived here are all for parallel-sided bodies. They could be modified for flow normal to a cylindrical surface. Resch (1967) has derived such an equation to characterize gas flow into cylindrical bolts assuming a constant coefficient.

The unsteady-state equation for gas flow into a cylinder may be derived from Darcy's law for an annulus. Recalling the derivation from Chap. 3.2, this may be written as

$$\frac{V}{t} = \frac{k_g 2\pi h \bar{P} dP}{P dR}.$$

When an unsteady-state equation is derived in a manner similar to that in Eq. (7.25) representing a parallel-sided body, the following result is obtained

$$\frac{\partial P^2}{\partial t} = \frac{k_g \bar{P}}{v_a}\frac{\partial}{\partial R}\left(R\frac{\partial P^2}{\partial R}\right).$$

Performing the differentiation,

$$\frac{\partial P^2}{\partial t} = \bar{D}_p\left(\frac{\partial^2 P^2}{\partial R^2} + \frac{1}{R}\frac{\partial P^2}{\partial R}\right). \tag{7.35}$$

(Unsteady-state gas flow in a cylinder)

7.3 Graphical and Analytical Solutions of Diffusion-Differential Equations with Constant Coefficients

The analytical solutions of the diffusion differential equations with constant coefficients [Eqs. (7.6), (7.16), (7.26), and (7.35)] are described by Crank (1956). These are based upon the following assumptions:

1. The diffusion coefficient is constant.
2. The initial value of temperature, moisture content, or pressure is uniform within the specimen.

3. The outside surfaces of the specimen immediately come to equilibrium with the uniform temperature, relative humidity, or pressure of the surroundings.
4. The transport of heat, moisture, or gas proceeds in a manner which is symmetrical in space throughout the body.

It should be emphasized that these solutions are accurate when the coefficient is nearly constant, as in the case of the transport of heat through wood. In the case of gas flow or moisture movement, finite-difference equations based upon Eqs. [(7.19), (7.30), (7.32), or (7.34)], should be used. It is possible, however, to obtain approximate results with these graphical or analytical solutions provided the corresponding diffusion coefficient does not vary appreciably over the range of values of pressure or moisture content. If the approximate Eqs. (7.16), (7.26), and (7.35) are solved in this manner, space and time averages of moisture content and pressure are required. In the case of moisture content, Eq. (6.9) may be used. For pressure, a simple arithmetic average of the initial and final average pressures is suggested

$$\bar{\bar{P}} = \frac{P_i + \bar{P}_f}{2}, \tag{7.36}$$

where $\bar{\bar{P}}$ = space and time average pressure, P_i = initial pressure, $\bar{P}_f$ = final average pressure.

It is convenient to simplify Eqs. (7.6), (7.16), and (7.26) to a general dimensionless form in which E or $\bar{E}$ is the dimensionless potential and τ the dimensionless time. The term E may be defined as:

$$E = \frac{T - T_i}{T_0 - T_i} = \frac{P^2 - P_i^2}{P_0^2 - P_i^2}, \tag{7.37a}$$

where E = fractional change in temperature or pressure squared at time t and at distance x or R from the center, T and P^2 = values of potential at time t and distance x or R from the center, T_i and P_i^2 = initial values of potential within the specimen at time zero, T_0 and P_0^2 = values of potential at external surfaces at all times.

Moisture potential is not included in the definition of E because it is usually expressed as an average value, even though a specimen may include a moisture gradient. However, if it is desired to calculate M at a point and at a given time, this may be similarly defined in terms of E.

Dimensionless time, τ, is defined as:

$$\tau = \frac{tD}{(L/2)^2} \qquad \tau = \frac{tD}{r^2}, \tag{7.37b}$$

(Parallel-sided) (Cylinder)

where t = time, s, D = diffusion coefficient (D_h, D_p, or D), cm²/s, L/2 = half thickness of parallel-sided body cm, r = radius of cylinder, cm.

When these substitutions are made, the general diffusion-differential equation may be written as

$$\frac{\partial E}{\partial \tau} = \frac{\partial^2 E}{\partial X}, \tag{7.38}$$

where X = dimensionless distance.

Similar equations may be written for average potentials within a body at time t. For this purpose $\bar{E}$ is defined as:

$$\bar{E} = \frac{M - M_i}{M_0 - M_i} = \frac{\bar{P}^2 - P_i^2}{P_0^2 - P_i^2}, \tag{7.39}$$

where $\bar{E}$ = fractional change in moisture content or pressure squared within the entire body, $\bar{P}^2$ = average pressure squared at time t.

The term $\bar{E}$ may be substituted into Eq. (7.38), resulting in an equation of identical form, but with different solutions because of the different values of $\bar{E}$ and E at a given time, although both these quantities have a value of zero at time zero and 1.0 at infinite time. It is clear from Eq. (7.39) that $\bar{E}$ is directly proportional to the change in the moisture content of a body at time t compared with that at infinity. It should be noted that this quantity has a different meaning when applied to gas flow because of the pressure-squared potential term. Therefore it is a measure of pressure squared and not quantity of gas (which is directly proportional to average pressure). A calculation of the fractional change in the quantity of gas may be accomplished in the following way:

$$F_g = \frac{\bar{P} - P_i}{P_0 - P_i}, \tag{7.40}$$

where F_g = fractional evacuation or filling with a gas, $\bar{P} \approx \sqrt{\bar{P}^2}$, $\bar{P}^2$ is calculated using Eq. (7.39). It is clear that $\sqrt{\bar{P}^2}$ is a root-mean-square pressure and not the same as the average pressure $\bar{P}$. The root-mean-square value is slightly higher than the average, but they are assumed equal in this calculation.

The value of F_g may be related to $\bar{E}$ by the following equations:

$$\bar{E} = 2F_g - F_g^2, \qquad \bar{E} = F_g^2 \tag{7.41}$$

(Evacuation, $P_0 = 0$)   (Filling, $P_i = 0$)

7.3.1 Solutions of Equations for Parallel-Sided Bodies

Crank (1956) presents an equation (his p. 45) from which values of potential throughout a body may be calculated. The constants have been combined in the form given below, which includes only the first two terms of a rapidly converging series.

$$E = 1.0 - 1.27 \exp(-2.47\ \tau) \cos[90°(2x/L)]$$
$$+ 0.425 \exp(-22.2\ \tau) \cos[270°(2x/L)]. \tag{7.42}$$

When the change in potential at the center of the specimen only is considered $(2x/L = \text{zero})$, Eq. (7.42) may be simplified to:

$$E = 1.0 - 1.27 \exp(-2.47\ \tau) + 0.425 \exp(-22.2\ \tau). \tag{7.43}$$

Values of E and τ calculated from Eq. (7.42) are plotted in Fig. 7.2 for $2x/L$ values of zero, 0.2, 0.4, 0.6, and 0.8. The value of zero corresponds to the center of the specimen and 1.0 to the outside surface. If τ is equal to or greater than 0.2, only the first exponential term of Eq. (7.42) is needed because of the rapid convergence of the series. Both exponential terms are significant if τ has a value between 0.07 and 0.2. Values of E vs τ at the center are plotted in Fig. 7.3. MacLean (1952) has presented several curves similar to Fig. 7.2, but especially adapted to heat flow into lumber of circular and rectangular cross section. These are discussed in Sect. 7.3.5.

The dimensionless potential $\bar{E}$ may be calculated from another equation on p. 45 of Crank (1956).

$$\bar{E} = 1.0 - 0.811 \exp(-2.47\,\tau), \tag{7.44}$$

where $\tau > 0.2$.

The value of $\bar{E}$ for short times may be determined as

$$\bar{E} = 1.13\sqrt{\tau}, \tag{7.45}$$

where $\tau < 0.3$.

The values of $\bar{E}$ determined from Eqs. (7.44) and (7.45) are plotted in Fig. 7.4.

Equation (7.45) may be rearranged in a form which is useful in the determination of moisture diffusion coefficients by the unsteady-state method

$$D = \frac{(\bar{E})^2 L^2}{5.10\,t}. \tag{7.46}$$

If the moisture content of a parallel-sided specimen is increased or decreased over a relatively small time interval ($\tau < 0.3$), the values of $\bar{E}$ at successive times may be calculated from the specimen weights. Assuming D constant over a small moisture-content range, there is a linear relationship between $(\bar{E})^2$ and t, or $\bar{E}$ and $\sqrt{t}$, as indicated in Fig. 7.5. The diffusion coefficient may then be calculated from the slope.

The surface resistance has been neglected in the steady-state methods of measuring the coefficient (Chap. 6.9) and in the unsteady-state method given above. Actually surface resistance may be a significant part of the total resistance to the transport of moisture between wood and air. It is due to the resistance of the air film adjacent to the wood and is reduced by an increase in air velocity. Choong and Skaar (1969, 1972) and Rosen (1978) have described the means of evaluating the relative effect of internal resistance, due to diffusion through the wood, and the external resistance. The method utilizes unsteady-state flow as represented by Eqs. (7.45) and (7.46). Several matched specimens of different thicknesses are equilibrated in a controlled environment and the half-times ($t_{0.5}$), or times to $\bar{E} = 0.5$, are measured. The equation used is

$$\tau_{0.5} = 0.2 + 0.7\,(D/S)(2/L), \tag{7.47}$$

where $\tau_{0.5}$ = dimensionless time calculated from $t_{0.5}$; 0.2 = value of $\tau_{0.5}$ for $\bar{E} = 0.5$ (Fig. 7.4) assuming no surface resistance, S = surface emission coefficient, cm/s, D = diffusion coefficient for internal resistance, assumed constant for matched specimens, cm^2/s.

It is evident from Eq. (7.47) that, if measurement of $\tau_{0.5}$ is made for two or more matched specimens, S may be determined because there are two unknowns, S and D, and two or more simultaneous equations. Rosen (1978) has rearranged Eq.

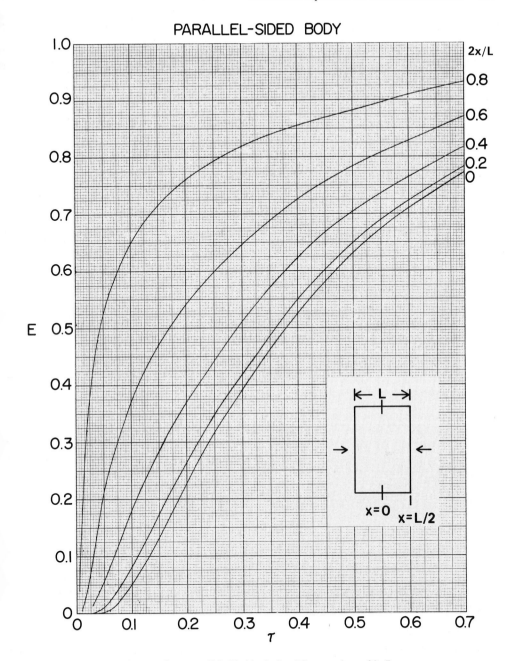

Fig. 7.2. Values of E vs. τ for a parallel-sided body for different values of $2x/L$

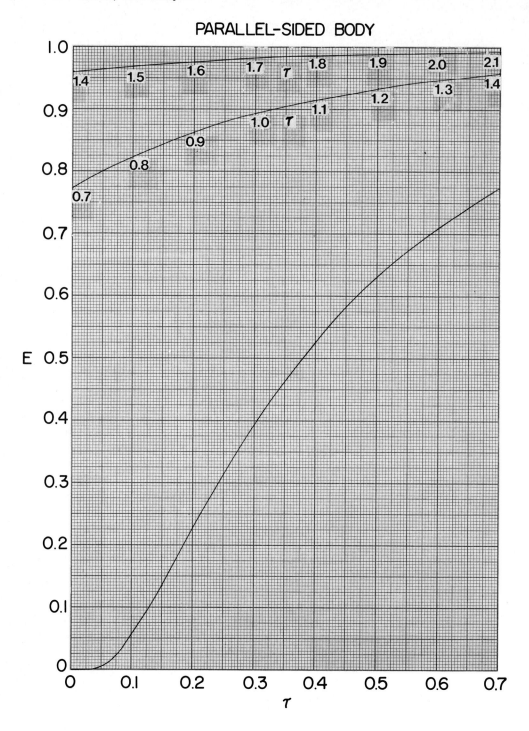

Fig. 7.3. Values of E vs. τ at the midpoint between two parallel surfaces ($2x/L = 0$)

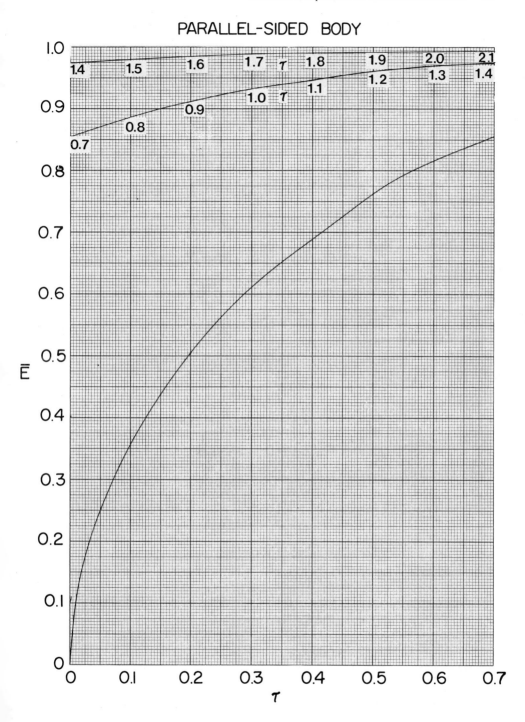

Fig. 7.4. Values of $\bar{E}$ vs. τ for a parallel-sided body

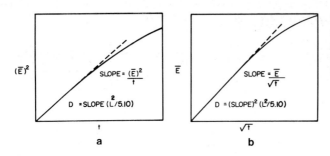

Fig. 7.5. Illustrating the determination of the average moisture diffusion coefficient by the unsteady-state method. a $\bar{E}^2$ vs t; b $\bar{E}$ vs $\sqrt{t}$

(7.47) by solving it for $\tau_{0.5}$.

$$\tau_{0.5} = \frac{0.2(L/2)^2}{D} + \frac{0.7(L/2)}{S}. \tag{7.48}$$

The first term of Eq. (7.48) represents the contribution of internal resistance and the second the external component. He has then calculated the fraction due to external resistance (F_{er}) as

$$F_{er} = \frac{0.7}{0.2 S(L/2)/D + 0.7}. \tag{7.49}$$

The results of Rosen's investigation disclosed that the rate of moisture transport in wood increased significantly with air velocity up to approximately 3 m/s with little change beyond this and that the fraction of external resistance increased as thickness decreased. In analyzing the results obtained for *Acer saccharinum* using Eq. (7.49), 15% of the total resistance was external for an air velocity 11.7 m/s and a thickness of 2.6 cm, while 89% was external resistance at a velocity of 0.43 m/s for a thickness of 0.7 cm. Similar results were obtained for *Juglans nigra*. It is apparent from this that external resistance is only negligible at air velocities in excess of 3 m/s and in relatively thick specimens. Since many of the measurements of D have been made with thin specimens at low air velocities, the values obtained would be expected to be significantly lower than the true value of D.

7.3.2 Solutions of Equations for Cylinders

The data presented in Fig. 7.6 for values of E at different ratios of R/r were taken from Fig. 5.3, p 67 of Crank (1956). The equation for calculating these values involves evaluating Bessel functons and so is not given here. The values of E corresponding to the center of the cylinder where R = 0 may be calculated from the following equation, which is plotted in Fig. 7.7.

$$E = 1.0 - 1.60 \exp(-5.76\,\tau) + 1.07 \exp(-30.6\,\tau), \tag{7.50}$$

where $\tau > 0.1$.

The series represented by Eq. (7.50) converges rapidly and only the first exponential term need be used for values of τ greater than 0.2.

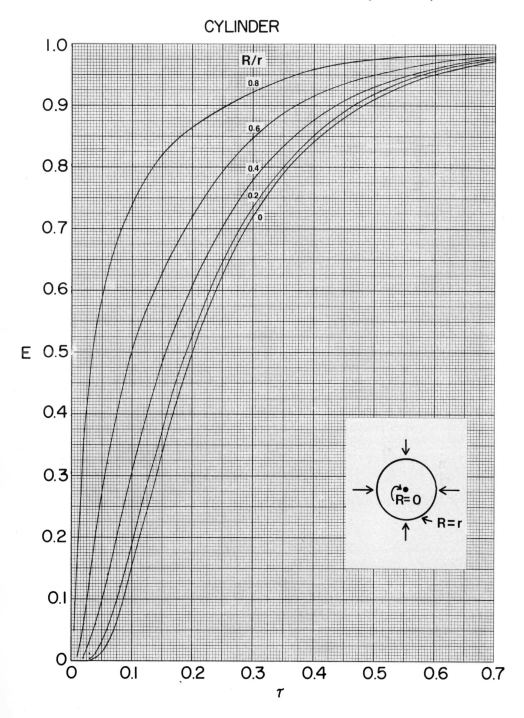

Fig. 7.6. Values of E vs. τ for a cylinder for different values of R/r

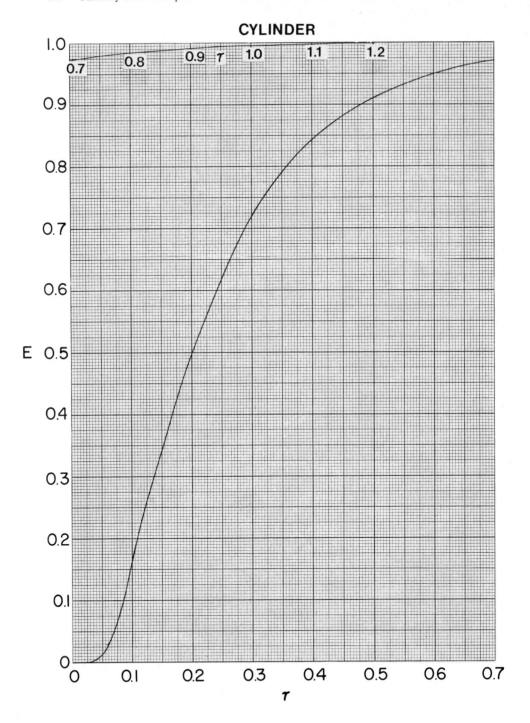

Fig. 7.7. Values of E vs. τ at the center of a cylinder (R = 0)

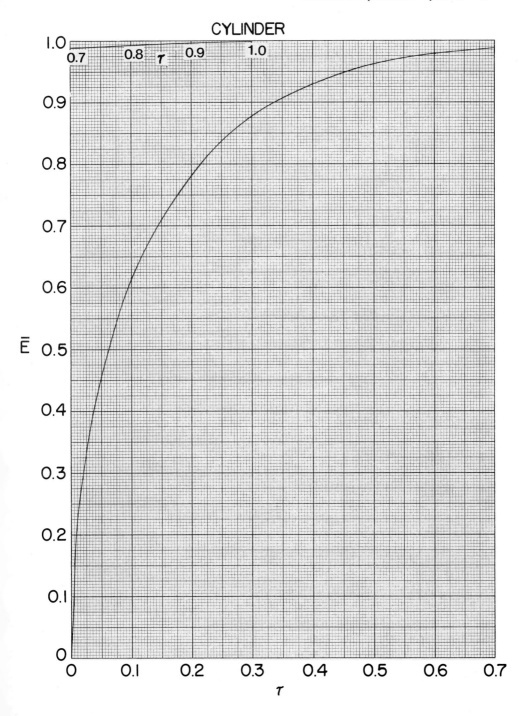

Fig. 7.8. Values of $\bar{E}$ vs. τ for a cylinder

Crank also presents equations for $\bar{E}$ where

$$\bar{E} = 1.0 - 0.693 \exp(-5.78\,\tau), \tag{7.51}$$

where $\tau \geq 0.1$.

For small times,

$$\bar{E} = 2.26\,\sqrt{\tau} - \tau. \tag{7.52}$$

The values of $\bar{E}$ calculated from Eqs. (7.51) and (7.52) are presented in Fig. 7.8.

7.3.3. Simultaneous Diffusion in Different Flow Directions

It is noted from Figs. 7.3, 7.4, 7.7, and 7.8 that the values of E and $\bar{E}$ for cylinders are much higher than those for plane-surfaced bodies for equal values of τ. This is reasonable, since the transport into a cylinder is directed radially all around the circumference, while that into a plane-surfaced body enters only from a pair of parallel surfaces, as indicated in Fig. 7.9. The cylindrical case is closely approximated by a square cross section as in Fig. 7.9c, where $L/2$ has the same value as r. It is also clear that E or $\bar{E}$ for a square cross section is slightly less than the corresponding value for a cylinder because of the larger area and longer average length of flow path to the center. The additional fractional change of potential possible at a given time before equilibrium is achieved can be represented by the quantity $(1-E)$ or $(1-\bar{E})$. This quantity is reduced by the contribution to flow through an additional pair of parallel surfaces. The value of $(1-E)$ or $(1-\bar{E})$ for an entire body or cross section is equal to the product of the corresponding values for the pairs of surfaces through which the flow occurs. Then

$$1-\bar{E} = (1-\bar{E}_T)(1-\bar{E}_R)(1-\bar{E}_L), \tag{7.53}$$

(Rectangular timber)

where the subscripts T, R, and L refer to tangential, radial, and longitudinal directions.

Equation (7.53) may also be used to calculate E for a rectangular timber by substitution of the appropriate values of E_T, E_R, and E_L.

If a square or rectangular cross section is considered, the longitudinal component in Eq. (7.53) is neglected for long lengths and the equation becomes

$$1-\bar{E} = (1-\bar{E}_T)(1-\bar{E}_R). \tag{7.54}$$

(Square or rectangular cross section)

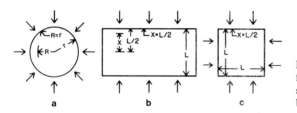

Fig. 7.9. Illustrating unsteady-state flow into bodies of different shapes: **a** cylindrical body; **b** parallel-sided body; **c** square cross section

If the diffusion coefficients are equal in the tangential and radial directions for a square cross section, the values of τ are equal, and $\bar{E}_T = \bar{E}_R$. Then Eq. (7.54) may be simplified to:

$$1-\bar{E} \text{ (square timber)} = (1-\bar{E}_T)^2$$

or

$$\bar{E} \text{ (square timber)} = 2\bar{E}_T - \bar{E}_T^2, \tag{7.55}$$

where $\bar{E}_T$ refers to flow in two transverse directions.

An *example* will illustrate the relationship between values of $\bar{E}$ for a cylindrical and square cross section of the same diameter with equal values of τ.

Assume $\tau = 0.3$,
$\bar{E}$ for a cylinder (Fig. 7.8) = 0.88,
$\bar{E}_T$ for a plane-surfaced body (Fig. 7.4) = 0.61,
$\bar{E}$ for a square cross section = $2(0.61) - 0.61^2 = 0.85$.

It is evident that the value of $\bar{E}$ for a cylinder is slightly higher than for a square cross section as expected, due to the decreased average distance and cross-sectional area into which flow occurs in a cylinder.

When a cylindrical solid is considered, the value of $\bar{E}$ or E may be calculated from the following equations:

$$1-\bar{E} \text{ (cylindrical solid)} = (1-\bar{E}_{cyl})(1-\bar{E}_L)$$

or

$$\bar{E} \text{ (cylindrical solid)} = \bar{E}_{cyl} + \bar{E}_L - \bar{E}_{cyl}\bar{E}_L. \tag{7.56}$$

Equations (7.53) through (7.56) are most useful in calculating values of E or $\bar{E}$ for the specimen from a given value of time as the independent variable. It is difficult to calculate an unknown time from a given value of $\bar{E}$ or E because of the complexity of the functions. This could be done by iteration.

7.3.4 Significance of Flow in Different Directions

It is frequently possible to simplify calculations by neglecting the flow in one or two directions of a three-dimensional solid. In the case of gas or liquid transport into a short, relatively thick cylinder it is generally possible to neglect transverse flow because of the high ratio of longitudinal to transverse permeabilities (usually > 10,000:1). Conversely it is usually possible to neglect longitudinal heat flow into a long, slender specimen because of the low ratio of longitudinal to transverse thermal conductivities (2.5:1). In regard to transverse water-vapor diffusion during drying, the flow in the direction of the width of a board is frequently negligible in comparison to that in the thickness direction.

In order to quantify these facts, a parallel-sided body is selected as an example with flow along two dimensions only. The flow along one of the dimensions is considered negligible if its contribution to the total is equal to or less than 10% of the

flow along the other dimension. To illustrate this, Eq. (7.54) may be rewritten as
$1-\bar{E} = (1-\bar{E}_1)(1-\bar{E}_2)$
or

$$\bar{E} = \bar{E}_1 + \bar{E}_2 - \bar{E}_1 \bar{E}_2 . \qquad (7.54\text{a})$$

Let the dimensions equal L_1 and L_2, corresponding to diffusion coefficients D_1 and D_2, with values of τ_1 and τ_2 corresponding to $\bar{E}_1$ and $\bar{E}_2$ in Eq. (7.54a). If the flow along L_2 is neglected, $\bar{E} = \bar{E}_1$. On the other hand, if the flow along L_2 contributes 10% of that along L_1 to the total, then $\bar{E} = 1.1 \bar{E}_1$. Substituting into Eq. (7.54a),

$$\bar{E}_2 = \frac{0.1}{1/\bar{E}_1 - 1}. \qquad (7.57\text{a})$$

If it is desired to calculate the value of $\bar{E}$, corresponding to negligible flow along the shorter dimension L_1, Eq. (7.57a) may be modified to

$$\bar{E}_1 = \frac{0.1}{1/\bar{E}_2 - 1}. \qquad (7.57\text{b})$$

The value of τ_1 and τ_2 both contain the same real time, t. Therefore, from Eq. (7.37b) it may be shown that

$$\frac{L_2}{L_1} = \sqrt{\frac{\tau_1 D_2}{\tau_2 D_1}}. \qquad (7.58)$$

It is clear from Eq. (7.58) that the ratio of τ's corresponding to $\bar{E}_1$ and $\bar{E}_2$ for a negligible contribution along L_2 or L_1 designates the locus of two lines on log-log graph paper which relate the length ratio (L_2/L_1) to the diffusion-coefficient ratio (D_2/D_1), with a slope of 0.5. Such plots appear in Fig. 7.10, derived to represent contributions to the value of $\bar{E}$ for a parallel-sided body. There are three pairs of such lines appearing in the graph because the values of τ_1 and τ_2 vary with the overall value of $\bar{E}$. The upper lines represent ratios of L_2/L_1 for neglecting the flow along L_2 for values of $\bar{E}_1$ of 0.3, 0.5, and 0.8, with $\bar{E}_2$ calculated from Eq. (7.57a), while the lower lines represent the ratios of L_2/L_1 for neglecting flow along L_1 for values of $\bar{E}_2$ of 0.3, 0.5, and 0.8, with $\bar{E}_1$ calculated from Eq. (7.57b).

An *example* will illustrate the derivation and use of Fig. 7.10. Assume $\bar{E}_1 = 0.5$. The corresponding $\bar{E}_2$ for a 10% contribution along L_2 is 0.10 from Eq. (7.57a). Reference to Fig. 7.4 and Eq. (7.45) gives corresponding values of τ_1 and τ_2 as 0.2 and 0.0078. The value of $\bar{E}$ for the entire specimen for these conditions is $0.1 + 0.5 - 0.1(0.5) = 0.55$ (Eq. 7.54a). A practical application of Fig. 7.10 is to the drying of lumber. Assume a board has dimensions of 1 in × 10 in × 96 in. First, to consider the importance of the width in the flow of moisture from the board, assume $D_2/D_1 = 1.0$ because the diffusion coefficients in the transverse directions are approximately equal. Let width be represented by L_2 and thickness by L_1, substituting into Eq. (7.58), $L_2/L_1 = \sqrt{0.2/0.0078} = 5.0$ which agrees with the value read in Fig. 7.10. Since the actual ratio of $L_2/L_1 = 10$, the contribution along the width is significantly less than 10% and is therefore negligible. A further determination may be made of the contribution along the length. Since the longitudinal moisture diffusion coefficient is much greater than the transverse, assume $D_2/D_1 = 80$. Using

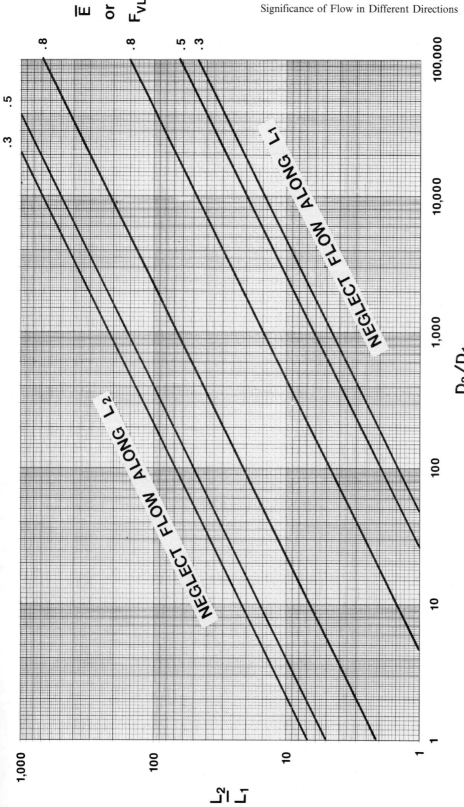

Fig. 7.10. Relative significance of flows in two directions in the determination of $\overline{E}$ or F_{VL} in a parallel-sided body; applicable to moisture diffusion and liquid flow

the same values of τ_1 and τ_2, $L_2/L_1 = \sqrt{(0.2/0.0078)}\,80 = 45$, which also appears in Fig. 7.10. Since the actual ratio of 96 exceeds this, the longitudinal contribution may be neglected in comparison with that of the thickness. It is further possible to calculate the ratio of L_2/L_1 for a negligible contribution along L_1. In this case $\bar{E}_2$ is assumed equal to 0.5 and $\bar{E}_1$ is calculated from Eq. (7.57b) with the result that $\tau_1 = 0.0078$ and $\tau_2 = 0.2$. Substitution into Eq. (7.58), $L_2/L_1 = \sqrt{(0.0078/0.2)(80)} = 1.75$ in agreement with Fig. 7.10. Therefore a specimen 1.75 inches long would have negligible flow along the thickness. Although such a situation would be rare in practice, it illustrates the use and derivation of the figure. Figure 7.11 is applicable to the moisture movement in a square or rectangular timber. In this case, if a ratio D_L/D_T of 100 is assumed, it is clear that longitudinal flow is negligible for L/d greater than 37 for $\bar{E}_T = 0.3$, 29 for $\bar{E}_T = 0.5$, and 13 for $\bar{E}_T = 0.8$.

Figure 7.12 may be used in the determination of E at the center of a rectangular cross section. In this case only one pair of lines is needed because the form of the curve in Fig. 7.3 is such that the criterion for neglect of flow is essentially independent of the value of E for the cross section. In applying this to heating a rectangular timber, flow along the longer dimension is negligible if L_2/L_1 exceeds 1.6 for equal values of diffusivity. In the comparison of flow in a transverse and longitudinal direction, $D_{hL}/D_{hT} = 2.5$ and the ratio L_2/L_1 is then 2.6. Therefore, if the length of the timber is at least 2.6 times the smaller dimension of the cross section, the longitudinal contribution may be neglected.

Figure 7.13 is most commonly applicable to heat transfer in a timber with a circular or square cross section. In this case, longitudinal heat flow may be neglected ($D_{hL}/D_{hT} = 2.5$) if the ratio of length to diameter ratio exceeds 1.9. On the other hand, transverse flow could be neglected with L/d less than 0.78.

When making calculations for gas flow in a cylindrical solid, values of $\bar{E}$ must be calculated from F_g, the fractional filling or evacuation with a gas using Eq. (7.41), because $\bar{E}$ is calculated from a potential of pressure squared. In this case values of F_g for the flow perpendicular to the cylindrical surface and along the length are calculated to correspond to a 10% contribution from either flow path. When this is done the value of ΔE needed to make a 10% contribution to F_g may be determined by differentiating Eq. (7.41) for evacuation.

$$\Delta \bar{E} = (2 - 2\,F_g)\,\Delta F_g. \tag{7.59}$$

Then with $\Delta F_g/F_g$ assumed equal to 0.1, the corresponding value of $\Delta \bar{E}/\bar{E}$ is calculated and substituted into Eqs. (7.57 a, b) in place of the decimal, 0.1, to determine the value of $\bar{E}$ or $\bar{E}_{cyl}$ corresponding to a negligible contribution. For example, assume F_g for cylindrical flow $= 0.5$ and $\Delta F_g = 0.05$. Then $\bar{E}_{cyl} = 0.75$ from Eq. (7.41) and $\Delta \bar{E}_{cyl} = 0.05$ from Eq. (7.59). The ratio $\Delta \bar{E}_{cyl}/\bar{E}_{cyl} = 0.05/0.75 = 0.067$. Equation (7.57a) may then be modified to calculate $\bar{E}_L$,

$$\bar{E}_L = \frac{0.067}{1/\bar{E}_{cyl} - 1}.$$

Then $\bar{E}_L = 0.201$ and $\tau_L = 0.032$ from Eq. (7.45) and $\tau_{cyl} = 0.175$ from Fig. 7.8. Assuming $D_{pL}/D_{pT} = 10{,}000$, $L/d = \sqrt{(0.175/0.032)\,10{,}000} = 234$ which appears in Fig. 7.14. Referring to the figure, assume a ratio of D_{pL}/D_{pT} of 20,000, a typical

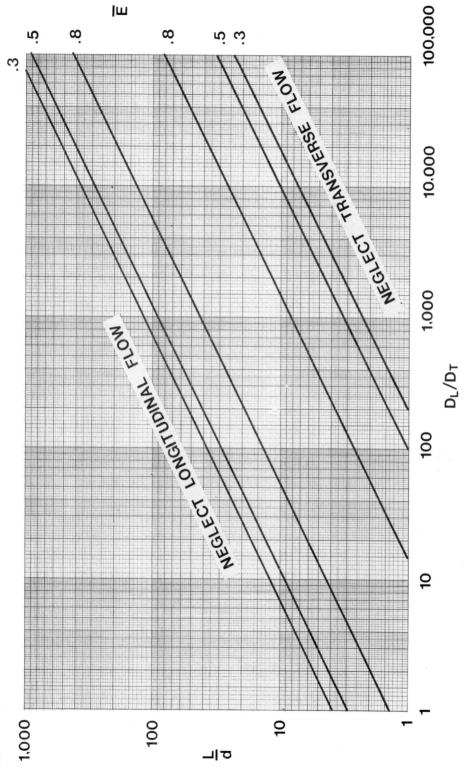

Fig. 7.11. Relative significance of longitudinal and transverse flow in the determination of $\bar{E}$ or F_{VL} for a cylindrical or square timber; applicable to moisture diffusion and liquid flow

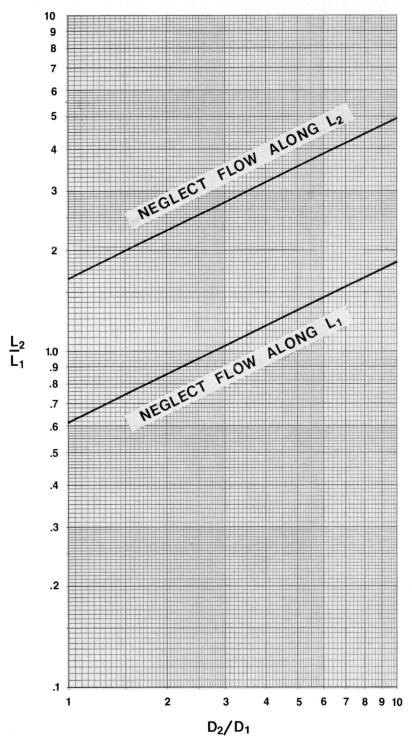

Fig. 7.12. Relative significance of flows in two directions in the determination of E at the center of a plane-surfaced body; applicable to heat transfer

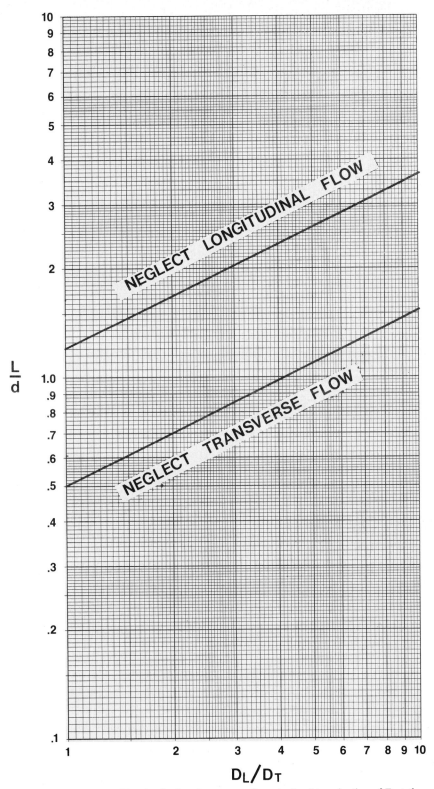

Fig. 7.13. Relative significance of longitudinal and transverse flows in the determination of E at the center of a cylindrical or square timber; applicable to heat flow

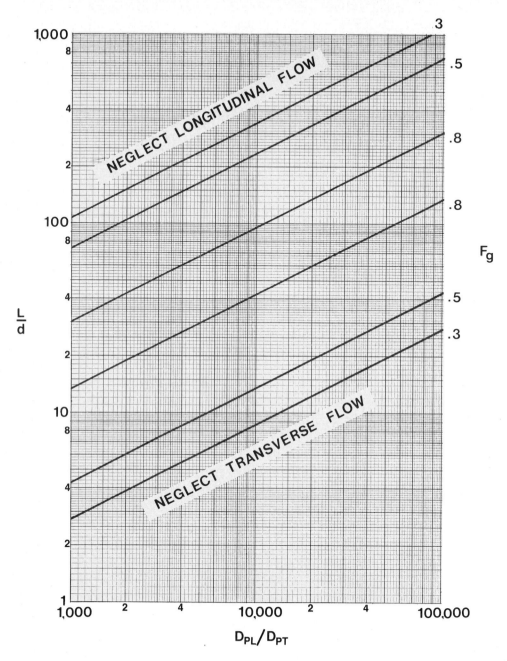

Fig. 7.14. Relative significance of longitudinal and transverse flows in the determination of F_g for gas flow in cylindrical or square timbers

value for softwoods. For $F_g = 0.5$, longitudinal flow may be neglected when L/d exceeds 330, on the other hand, transverse flow may be neglected when L/d is less than 19.

7.3.5 Special Considerations Relating to the Heating of Wood

MacLean (1952) has published several special graphical solutions of the unsteady-state differential equation applied to the heating of wood. These are useful for obtaining a rapid solution, and some of them are reproduced here. Figure 7.15 may be used to determine the temperature at any point within a long 10-inch-diameter pole having an initial temperature (T_i) of 60 °F, an outside temperature (T_0) of 200 °F, and a thermal diffusion coefficient of 1.62×10^{-3} cm²/s, corresponding approximately to that for wood with a specific gravity of 0.5 and a moisture content of 10%. It is clear that Fig. 7.15 presents the same information as Fig. 7.6, but in specialized form. The values of time may be corrected for different diffusion coefficients or diameters, since the time is inversely proportional to the diffusion coefficient and directly proportional to the square of the diameter. The following equation may be used to correct time values obtained from Fig. 7.15 to those corresponding with different diameters and thermal diffusion coefficients:

$$t = t' \left(\frac{d}{10}\right)^2 \frac{1.62 \times 10^{-3}}{D_{hT}} \tag{7.60}$$

or, solving for t',

$$t' = t \left(\frac{10}{d}\right)^2 \frac{D_{hT}}{1.62 \times 10^{-3}}, \tag{7.61}$$

where t = corrected time, t' = time as read from Fig. 7.15, d = diameter of pole, in inches, D_{hT} = transverse thermal diffusion coefficient of wood in pole, cm²/s.

Figure 7.15 may also be used for different values of T_0 and T_i because the change in the temperature of the wood is proportional to the difference between the outside and initial temperatures ($T_0 - T_i$). The following relationship may be used:

$$T = T_i + \frac{(T' - 60°)(T_0 - T_i)}{140°} \tag{7.62}$$

or, solving for T',

$$T' = \frac{140°(T - T_i)}{T_0 - T_i} + 60°, \tag{7.63}$$

where T = corrected temperature at proportional distance F/r from outside surface of pole, T_i = initial temperature of pole, T_0 = outside temperature of pole, T' = temperature of pole at proportional distance F/r as read from Fig. 7.15, 140° = difference between outside and initial temperatures for which Fig. 7.15 is calculated, 60° = initial temperature for which Fig. 7.15 is calculated.

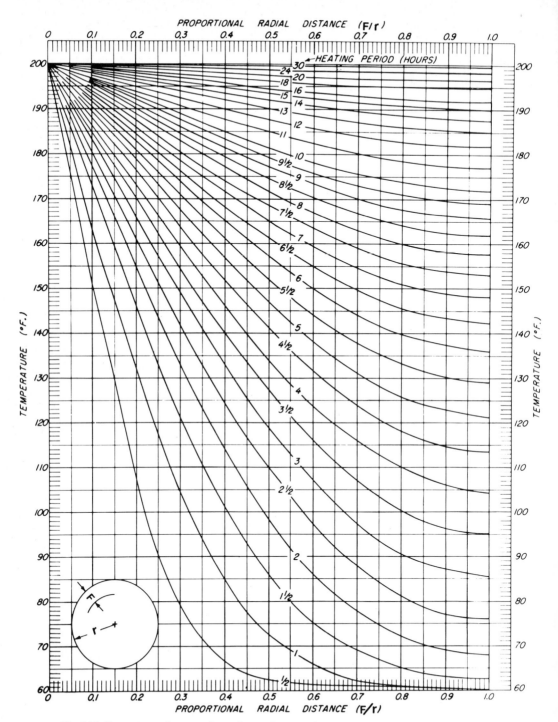

Fig. 7.15. Temperatures between the surface and center of a 10-inch-diameter long pole after various heating periods. Distance from surface is expressed as a proportion of radius (F/r). $D_{hT} = 1.62 \times 10^{-3}$ cm²/s; $T_i = 60$ °F; $T_o = 200$ °F. (MacLean 1952)

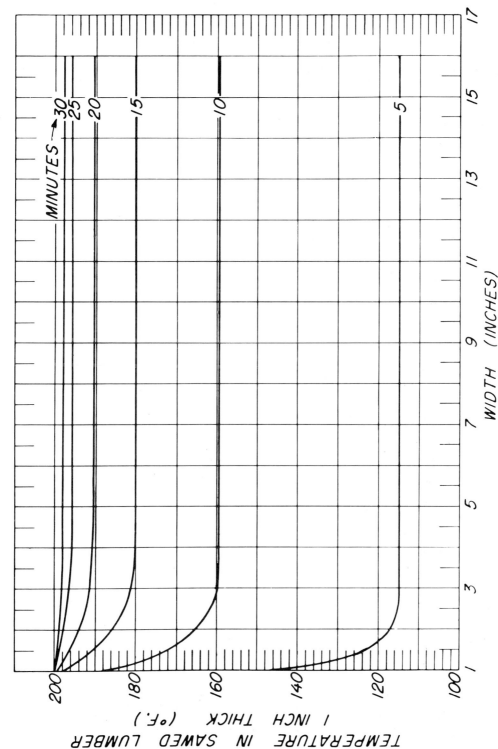

Fig. 7.16. Temperatures at the center of sawed lumber one inch thick after various heating periods. $D_{hT} = 1.62 \times 10^{-3}$ cm^2/s; $T_i = 60\,°F$; $T_0 = 200\,°F$. (MacLean 1952)

The center temperatures of 1-inch and 2-inch lumber may be determined from Figs. 7.16 and 7.17 and that of long rectangular timbers from Fig. 7.18. The temperature midway between the surface and center of rectangular timbers may be determined using Fig. 7.19. The flow in both the tangential and radial directions is accounted for in these figures, making them considerably easier to use than Fig. 7.2 for a parallel-sided body, which must be applied to both directions using the procedure outlined in Sect. 7.3.3.

MacLean also presents data to show that green wood in steam has a thermal diffusion coefficient approximately 22% higher than in creosote, corresponding with an 18% reduction in heating time. The values of thermal diffusion coefficients for seasoned wood in creosote are in good agreement with those calculated using Eq. (7.9.).

It is clear from Eq. (7.9) that the thermal diffusion coefficient of wood is affected by the thermal conductivity, the specific gravity, and the moisture content. Below the fiber saturation point the thermal conductivity increases with moisture content, while specific gravity decreases slightly, but the specific heat term increases appreciably. The net result is a decrease in thermal diffusion coefficient with an increase in moisture content below the fiber saturation point. Above the fiber saturation point the increases in thermal conductivity and specific heat almost balance, so there is little change in the thermal diffusion coefficient. Referring to Eq. (7.10), it is clear that the thermal conductivity increases with the specific gravity but not as rapidly, resulting in a decrease in thermal diffusion coefficient with an increase in specific gravity, as depicted in Fig. 7.20. In this case, the moisture content is constant. The effect of specific gravity on the diffusion coefficient is much less than its effect upon the thermal conductivity, since the changes in K_g and G tend to cancel each other out in the former instance. The lower diffusivities for denser woods are reasonable because more time is required to heat the greater amount of cell-wall substance. Even though the transfer of heat is more rapid in denser wood, this factor is overriden by the greater mass of material to be heated.

7.4 Relative Values of Diffusion Coefficients

In unsteady-state transport it is evident that hydrodynamic flow of gases takes place relatively rapidly, heating of wood in large sections requires several hours, and the drying of wood may require from a few days to serveral months depending on the thickness, specific gravity, and drying conditions. Also, the time required is a function of the direction relative to the fiber axis as summarized below:

Heat: $D_{hL}/D_{hT} = 2.5$,
Hydrodynamic gas flow: D_{pL}/D_{pT} 300 to 100,000,
Moisture diffusion: $D_L/D_T = $ 2 to 4 at M = 30%
                                50 to 100 at M = 5%.

The overall range of values of coefficients for wood are presented in Table 7.1. The possible variation in the values of thermal diffusion coefficients for all woods and conditions is relatively small, with the ratio of the maximum to minimum value

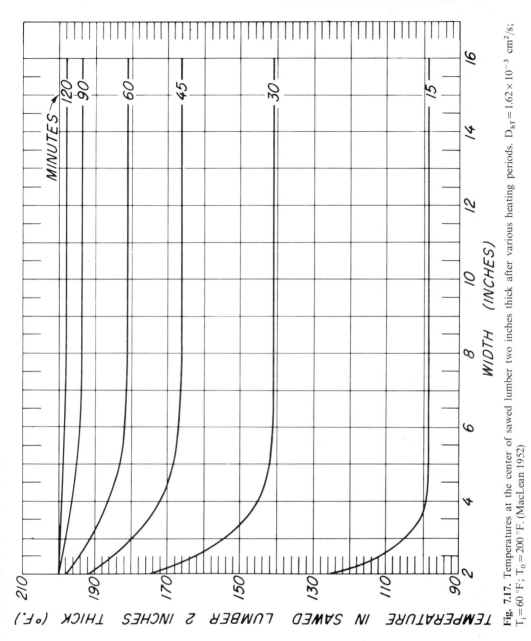

Fig. 7.17. Temperatures at the center of sawed lumber two inches thick after various heating periods. $D_{hT} = 1.62 \times 10^{-3}$ cm²/s; $T_i = 60\,°F$; $T_0 = 200\,°F$. (MacLean 1952)

being 2:1 or less in a given grain direction. The factors influencing thermal diffusivity are thermal conductivity, specific gravity, and moisture content, and the effects of these were discussed in the previous section.

The range of values of moisture diffusion coefficients is much greater than that of the thermal diffusion coefficients because the values of both D_{BT} and D_v vary widely with temperature and moisture content (Fig. 6.5). Also, the variations with

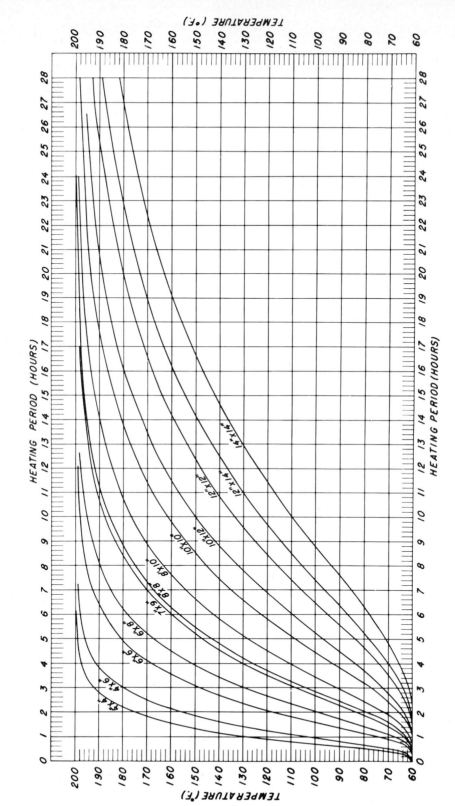

Fig. 7.18. Temperatures at the center of sawed timbers of various cross-sectional dimensions heated at 200 °F (T_0) for various periods of time. $D_{hT} = 1.62 \times 10^{-3}$ cm^2/s; $T_i = 60$ °F. (MacLean 1952)

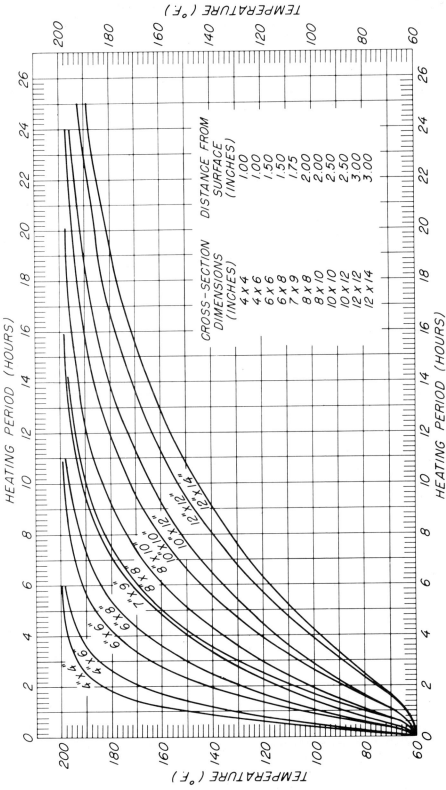

Fig. 7.19. Temperatures midway between the surface and center of sawed timbers of various cross-sectional dimensions (distance measured on the short axis when width is greater than thickness) for different periods of time. $D_{hT} = 1.62 \times 10^{-3}$ cm^2/s; $T_i = 60\,°F$; $T_0 = 200\,°F$. (MacLean 1952)

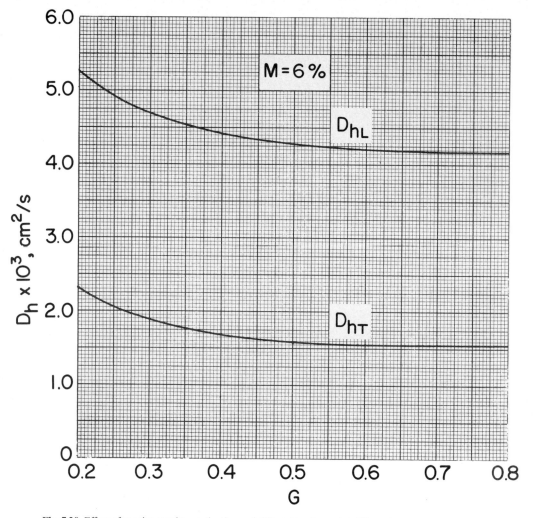

Fig. 7.20. Effect of species gravity on the thermal diffusivity of wood at 6% moisture content

changes in specific gravity are greater than those in the thermal diffusion coefficients because of the greater relative conductivity of the lumens compared with the cell walls. The variations in D_L and D_T with specific gravity at an average moisture content of 20% and at 40 °C are presented in Fig. 7.21.

7.5 Retention

Retention is a measure of the concentration of preservative or other liquid in wood, and it is expressed in pounds of preservative per cubic foot of wood in the North American wood-preserving industry while in S.I. units it would be kilograms per

Table 7.1. Overall range of values of coefficients for wood

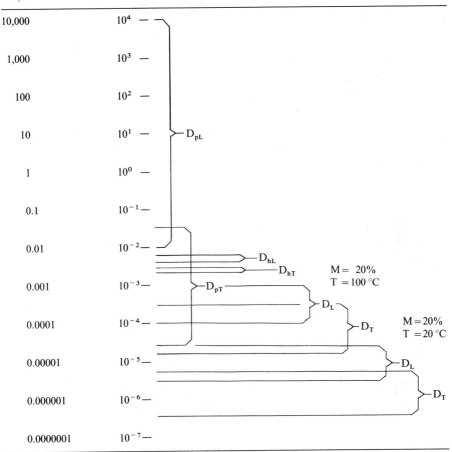

cubic meter of wood. Therefore its determination requires measurement of the volume of the wood to be treated. Conversion factors between these systems are listed below:

$$1 \text{ g/cm}^3 = 62.4 \text{ lb/ft}^3 = 1{,}000 \text{ kg/m}^3,$$
$$1 \text{ lb/ft}^3 = 0.0161 \text{ g/cm}^3 = 16.1 \text{ kg/m}^3.$$

Retention may be calculated using the following equation:

$$\text{Retention} = \frac{W_L}{V}, \tag{7.64}$$

where W_L = mass of liquid preservative, g, kg, or lb, V = volume of wood, cm³, m³ or ft³.

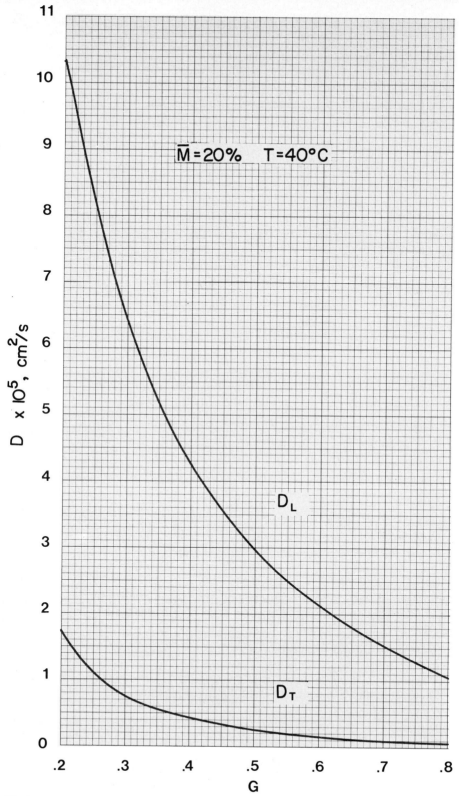

Fig. 7.21. Influence of specific gravity on the bound-water diffusion coefficient of wood at a moisture content of 20% at 40 °C; $D_v = 2.2 \times 10^{-5}$ cm²/s; $D_{BT} = 2.7 \times 10^{-7}$ cm²/s

Retention may also be expressed as the fraction of voids filled by liquid. This may be calculated by dividing the volume of the retained liquid by the volume of the voids in the wood.

$$F_{vL} = \frac{V_L}{V v_a}, \qquad (7.65)$$

where F_{vL} = fraction of voids filled by liquid, V_L = volume of liquid, v_a = porosity of wood, V = volume of wood.

The volume of the liquid may be calculated from its mass and density. Then

$$F_{vL} = \frac{W_L}{\varrho_L V v_a}, \qquad (7.66)$$

where ϱ_L = density of liquid, g/cm^3.

It is possible to convert fractional volumetric retention to the more practical units of lb/ft^3, g/cm^3, or kg/m^3 by combining Eqs. (7.64) and (7.66).

Retention, lb/ft^3 = 62.4 $F_{vL} \varrho_L v_a$, (7.67a)

Retention, g/cm^3 = $F_{vL} \varrho_L v_a$, (7.67b)

Retention, kg/m^3 = 1,000 $F_{vL} \varrho_L v_a$. (7.67c)

The maximum possible retention for a given wood specimen may be calculated by letting F_{vL} equal unity.

Maximum retention, lb/ft^3 = 62.4 $\varrho_L v_a$. (7.68)

As an *example*, assume a specimen with a specific gravity of 0.5, a moisture content of 6%, and a porosity of 0.63 (Fig. 5.9). The maximum possible retention of creosote (ϱ_L = 1.05 g/cm^3) is approximately 41 lb/ft^3, 0.66 g/cm^3, or 660 kg/m^3.

7.6 Unsteady-State Transport of Liquids

The flow of a liquid into a porous body is a different phenomenon than the diffusion of a gas because the liquid enters as a front with liquid–gas interfaces in the capillaries. If the uniform-parallel-circular capillary model is assumed, the front has a uniform penetration in all the capillaries. The equations to be discussed will be derived from Darcy's law and therefore include the assumptions upon which this law is based. Generally capillary forces are neglected and zero initial pressure is assumed. Therefore, air is evacuated from the specimens before the liquid is brought in. Such a method of impregnation is called a full-cell or Bethell process in the wood-preservation industry (Henry 1973).

7.6.1 Parallel-Sided Bodies, Permeability Assumed Constant with Length

Darcy's law for liquids may be written in derivative form as:

$$\frac{dV}{dt} = \frac{KA\Delta P}{\eta x}, \qquad (7.69)$$

where x = penetration, $dV = v_a A\, dx$, K = specific permeability.

Then,

$$\int_0^x x\,dx = \frac{K\Delta P}{v_a\eta}\int_0^t dt\,;$$

performing the integration,

$$x = \sqrt{\frac{2K\Delta Pt}{v_a\eta}}. \tag{7.70}$$

(Parallel-sided body)

Let

$$\tau = \frac{Kt\Delta P}{\eta v_a(L/2)^2},$$

then

$$x = L\sqrt{0.5\,\tau}, \tag{7.71}$$

(Parallel-sided body)

where L = thickness of body, with the maximum penetration being $L/2$ for the liquid entering from two opposite surfaces.

The volumetric absorption for entry on two opposite faces is:

$$V = 2v_a A x$$

or

$$V = 2v_a A \sqrt{\frac{2K\Delta Pt}{v_a\eta}}. \tag{7.72}$$

This may be expressed on a fractional volumetric retention basis. Since $F_{VL} = 2x/L$, assuming total filling of the voids from the surface to the liquid–gas interface.

Then

$$F_{VL} = \frac{2}{L}\sqrt{\frac{2K\Delta Pt}{v_a\eta}}. \tag{7.73}$$

(Parallel-sided body)

In terms of τ,

$$F_{VL} = \sqrt{2\tau}. \tag{7.74}$$

(Parallel-sided body)

A plot of F_{VL} vs τ is revealed in Fig. 7.22. There is some analogy between this and the plot of $\bar{E}$ vs τ (Fig. 7.4) for the diffusion equations, but the times are less in the former with $\tau = 0.125$ for $F_{VL} = 0.5$ compared with 0.2 in the latter for $\bar{E} = 0.5$. If it is desired to calculate an $\bar{E}$ value from liquid flow data, it is necessary to put the result on a basis of available void volume rather than the total void volume

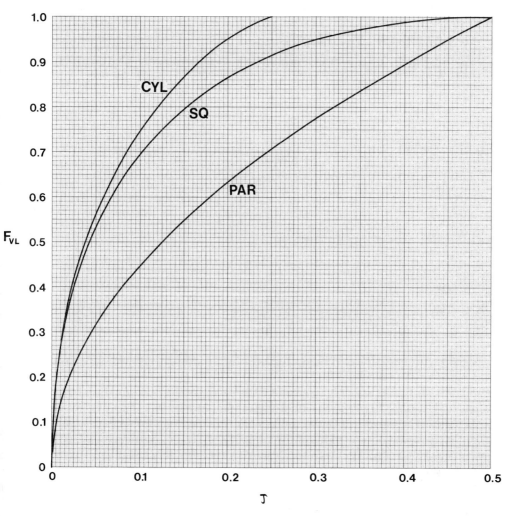

Fig. 7.22. Values of F_{VL} vs. τ for liquid flow into parallel-sided bodies *(PAR)* calculated from Eq. (7.74), square cross sections *(SQ)* calculated from Eq. (7.88), and cylindrical cross sections *(CYL)* calculated from Eq. (7.84). The value of τ is taken as $(4 K t \Delta P)/(v_a \eta L^2)$ for a parallel-sided body or square cross section and as $(K t \Delta P)/(v_a \eta r^2)$ for a cylinder

fraction, v_a. This may be done by introducing a factor to account for nonhomogeneity of the flow properties.

$$\bar{E} = \sqrt{2 \tau / f}, \tag{7.75}$$

where f = fraction of cross section of wood available for liquid transport.

The equations derived above predict that the penetration and volumetric retention are proportional to square root of time. This relationship was confirmed by Petty (1975) for *Pinus sylvestris* and later (1978 b) for several permeable softwoods. Petty (1975) also found evidence of two parallel zones of widely differing permeabilities in *Pinus sylvestris* because the plot of absorption vs $\sqrt{t}$ consisted of two

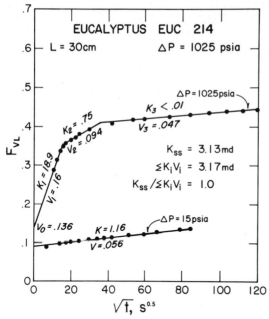

Fig. 7.23. Plot of F_{VL} vs. square root of time for 30-cm-long specimen of *Eucalyptus deglupta* treated at atmospheric pressure until absorption stopped and then at 1,025 lb/in². Three parallel zones of progressively decreasing permeabilities are evident. The permeability during the atmospheric-pressure period was much less than during the high-pressure period, probably due to the ability of the high pressure to overcome capillary forces in excess of one atmosphere. Permeabilities are given in millidarcys (md). K_{ss} = measured steady-state permeability, $\Sigma K_i V_i$ was calculated from Eq. (7.76). (Gonzalez and Siau 1978)

straight segments of different slopes followed by a third segment with zero slope. The portion with steeper slope was attributed to latewood and the other to earlywood. Similar results were obtained by Gonzalez and Siau (1978) for *Fagus grandifolia* and *Eucalyptus deglupta* with a typical plot in Fig. 7.23. This plot reveals three zones. The individual permeabilities and volume fractions were calculated from the slopes and intersections of the straight lines. In this particular case there was very close agreement between the steady-state permeability and that calculated from the sum of the permeabilities and volume fractions corresponding with the three zones using the following relationship:

$$K = \sum_1^n K_i v_i = K_1 v_1 + K_2 v_2 + \ldots, \tag{7.76}$$

where K = permeability of the specimen, K_i and v_i are permeability and volume fraction of component i, n = no. of components.

Petty (1978b) has refined Eq. (7.73) for use with initial pressure above vacuum and a final pressure of 1 atm. He also added a term to account for the combined effects of capillary forces, vapor pressure of the liquid, and the pressure of the dissolved air. Relatively good agreements were obtained between the measured and predicted fractional volumetric retentions.

7.6.2 Parallel-Sided Bodies with Permeability Decreasing with Length (Bramhall Model)

Permeability is usually independent of specimen length in woods of relatively high permeability; however Sebastian et al. (1965), Bramhall (1971), and Siau (1972)

(Fig. 3.12) have observed that permeability decreases with length in woods of low permeability (less than 2 darcys) with the effect tending to become greater as the permeability decreases. Bramhall observed a linear relationship between the logarithm of the permeability and the length of specimens of *Pseudotsuga menziesii* with lengths between 0.5 and 3.5 cm. Bramhall explained this by an exponential decrease in the conductive area of the cross section resulting in a modified Darcy's law equation of the form:

$$K_0 = \frac{VL\eta}{tAe^{-bL}\Delta P}, \qquad (7.77)$$

where K_0 = intercept permeability at zero length (a constant), b = exponential coefficient obtained from the slope of a plot of ln K vs L.

The permeability at any given length is calculated from the Darcy's law equation as:

$$K = \frac{VL\eta}{tA\Delta P}.$$

Equation (7.77) may be written in derivative form as

$$\frac{dV}{dt} = \frac{K_0 A e^{-bx}\Delta P}{x\eta}. \qquad (7.78)$$

Since $dV = A v_a e^{-bx} dx$, the exponential term appears on both sides and divides out. Therefore the penetration is the same as that for constant permeability calculated from Eq. (7.70).

The total volumetric absorption, V, may be obtained by integration:

$$\int_0^V dV = 2v_a A \int_0^x e^{-bx} dx,$$

where the factor of 2 is inserted to account for penetration from both ends.
Performing the integration,

$$V = \frac{2v_a A}{b}(1 - e^{-bx}).$$

Letting $x = c\sqrt{t}$, where c = combined constants from Eq. (7.70), and solving for F_{VL},

$$F_{VL} = \frac{2}{Lb}(1 - e^{-bc\sqrt{t}}). \qquad (7.79)$$

It is clear that, although the penetration is proportional to $\sqrt{t}$, the fractional volumetric retention is not. This has been verified by Petty (1975) in a measurement of the longitudinal absorption of petroleum distillate of the relatively impermeable *Pseudotsuga menziesii* heartwood.

7.6.3 Cylindrical Specimens

Darcy's law for an annulus may be written as:

$$\frac{V}{t} = \frac{2\pi Kh\Delta P}{\eta \ln r/R}, \tag{7.80}$$

where h = length of annulus, r = outside radius, R = variable inside radius.
 Then $dV = v_a 2\pi h R dR$.
 The total volumetric absorption is then:

$$2\pi h v_a \int_R^r R dR = \pi h v_a (r^2 - R^2).$$

Since the volumetric absorption at full saturation is

$$\pi r^2 h v_a,$$

the fractional volumetric retention may be calculated as

$$F_{VL} = \frac{r^2 - R^2}{r^2}. \tag{7.81}$$

In order to find a relationship between F_{VL} and time it is necessary to integrate Eq. (7.80). Separating the variables,

$$\int_r^R -\ln(R/r) R dR = \frac{K\Delta P}{v_a \eta} \int_0^t dt. \tag{7.82}$$

Integration of Eq. (7.82) by parts yields:

$$-R^2/2 \ln r/R + (r^2 - R^2)/4 = K \Delta P t/(v_a \eta).$$

Then by substitution of Eq. (7.81) and dividing through by r^2,

$$-0.5 R^2/r^2 \ln r/R + 0.25 F_{VL} = \frac{K \Delta P t}{v_a \eta r^2}.$$

Let

$$\tau = \frac{K\Delta P t}{v_a \eta r^2},$$

then

$$\tau = 0.25 F_{VL} - 0.5 (R^2/r^2) \ln r/R. \tag{7.83}$$

(Cylinder)
 By substitution from Eq. (7.81),

$$\tau = 0.25 F_{VL} + 0.5 (1-F_{VL}) \ln (1-F_{VL})^{0.5}. \tag{7.84}$$

(Cylinder)
 Solving Eq. (7.83) for F_{VL},

$$F_{VL} = 4\tau + 2 (R^2/r^2) \ln (r/R). \tag{7.85}$$

Equation (7.84) is also plotted in Fig. 7.22 where the values of τ are significantly less than those for a parallel-sided body as expected.

The penetration in a cylinder is equal to (r–R) and may be calculated as

$$r-R = r\,(1-\sqrt{1-F_{VL}}). \tag{7.86}$$

It is clear from Eqs. (7.85) and (7.86) that neither F_{VL} nor penetration are proportional to $\sqrt{t}$ as in a parallel-sided body.

7.6.4 Square and Rectangular Specimens

For a square timber, the fractional volumetric retention may be calculated from that for a parallel-sided body [Eq. (7.74)] by a relationship similar to Eq. (7.55).

$$F_{VL\,(square\ timber)} = 2\,F_{VL} - F_{VL}^{2}. \tag{7.87}$$

Then, by substitution of τ from Eq. (7.74), assuming equal coefficients in the two directions,

$$F_{VL} = \sqrt{8\,\tau} - 2\,\tau. \tag{7.88}$$

(Square timber)

The penetration will be the same as that for a parallel-sided body as calculated from Eq. (7.70).

Equation (7.88) is plotted in Fig. 7.22 where it is evident that more time is required to penetrate a square than a circular cross section having the same cross-sectional dimension.

A relationship for the retention of a rectangular cross section with equal coefficients may be derived from Eq. (7.88) with the following result:

$$F_{VL} = (1 + L_1/L_2)\sqrt{2\,\tau_1} - 2\,(L_1/L_2)\,\tau_1, \tag{7.89}$$

(Rectangular cross section)
where τ_1 = dimensionless time for small side of the rectangle with length L_1, L_2 = long dimension of rectangle.

Figure 7.22 indicates a greater rate of flow through a square or circular cross section than through two parallel surfaces only. The fraction of additional flow which will result in a rectangular solid when both the narrow (L_1) and wide (L_2) dimensions are considered is plotted in Fig. 7.24 for various L_2/L_1 ratios. In the case of $L_2/L_1 = 1.0$ corresponding to a square cross section, the flow rate is doubled at very low values of τ (and penetration) as expected because there is very little overlapping between the fluxes from the two directions. At $\tau = 0.5$, corresponding to complete filling, the fractional increase is zero. Thus the two curves in Fig. 7.22 come together at $F_{VL} = 1.0$ or $\tau = 0.5$.

The relative significance of flows along the two dimensions of a rectangular cross section may be determined from Fig. 7.10 because of the close similarity between Eqs. (7.45) and (7.74). In addition, the criteria for neglecting either longitudinal or transverse flow in a cylindrical or square timber may be determined from Fig. 7.11.

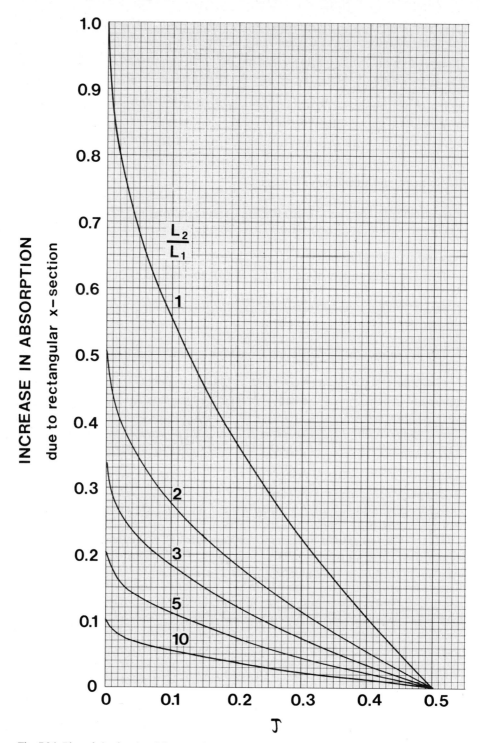

Fig. 7.24. Plot of the fractional increase in absorption of rectangular cross sections with various ratios of L_2/L_1 over that of two infinitely wide parallel surfaces as a function of τ. The calculations were made from Eqs. (7.74) and (7.89)

7.7 Unsteady-State Transport of Moisture under Nonisothermal Conditions

Nonisothermal moisture movement in the steady state was discussed in detail in Chap. 6.8, which will be referred to in the derivations proposed to characterize unsteady-state nonisothermal transport (Siau 1983b). Two steady-state equations were derived: one based upon a gradient of activated moisture molecules and the other upon chemical potential. The former wil be discussed first. Recalling Eq. (6.33) and rearranging it,

$$J = -\frac{G\varrho_w}{100}\left[D_T\left(\frac{M}{RT+70\,M}\right)\left(\frac{9{,}200-70\,M}{T}\right)\frac{dT}{dx} + D_T\frac{dM}{dx}\right], \qquad (7.90)$$

where D_T is defined by Eqs. (6.14) and (6.22) as

$$D_T = \frac{0.07\,\exp\left[-(9{,}200-70\,M)/RT\right]}{(1-a^2)(1-a)}.$$

A derivation is then employed based upon the moisture gained by a thin element of a parallel-sided body, similar to that for Eq. (7.18), but with the additional differentiation required to account for the effects of temperature and moisture content upon the diffusion coefficient and upon the other terms in Eq. (7.90). Let

$$\Phi = D_T\left(\frac{M}{RT+70\,M}\right)\left(\frac{9{,}200-70\,M}{T}\right),$$

then Eq. (7.90) may be simplified to

$$J = -\frac{G\varrho_w}{100}\left(\Phi\frac{dT}{dx} + D_T\frac{dM}{dx}\right).$$

The resulting unsteady-state equation used in place of Eq. (7.18) is then

$$\frac{\partial M}{\partial t} = \frac{\partial}{\partial x}\left[\Phi\frac{dT}{dx} + D_T\frac{dM}{dx}\right].$$

Performing the differentiations

$$\left(\frac{\partial M}{\partial t}\right)_x = \Phi\left(\frac{\partial^2 T}{\partial x^2}\right)_t + \left[\left(\frac{\partial \Phi}{\partial T}\right)_{t,M}\left(\frac{\partial T}{\partial x}\right)_t + \left(\frac{\partial \Phi}{\partial M}\right)_{t,T}\left(\frac{\partial M}{\partial x}\right)_t\right]\left(\frac{\partial T}{\partial x}\right)_t + D_T\left(\frac{\partial^2 M}{\partial x^2}\right)_t$$

$$+ \left[\left(\frac{\partial D_T}{\partial M}\right)_{t,T}\left(\frac{\partial M}{\partial x}\right)_t + \left(\frac{\partial D_T}{\partial T}\right)_{t,M}\left(\frac{\partial T}{\partial x}\right)_t\right]\left(\frac{\partial M}{\partial x}\right)_t, \qquad (7.91)$$

(Gradient of activated molecules)

where the differentiation of the variable Φ is performed as the product of three variables.

When chemical potential is considered as the driving force, a similar derivation may be used. Equation (6.45) may be rearranged as

$$J = -\frac{G\varrho_w}{100}\left[D_T\left(\frac{H}{RT}\right)\left(\frac{\partial M}{\partial H}\right)\left(\frac{\partial \mu_1^0}{\partial T} + \frac{E_L}{T} + R\ln H/100\right)\frac{dT}{dx} + D_T\frac{dM}{dx}\right]. \quad (7.92)$$

Let

$$\Theta = D_T\left(\frac{H}{RT}\right)\left(\frac{\partial M}{\partial H}\right)_T\left(\frac{\partial \mu_1^0}{\partial T} + \frac{E_L}{T} + R\ln H/100\right).$$

The resulting unsteady-state equation then may be written as

$$\left(\frac{\partial M}{\partial t}\right)_x = \Theta\left(\frac{\partial^2 T}{\partial x^2}\right)_t + \left[\left(\frac{\partial \Theta}{\partial T}\right)_{t,M}\left(\frac{\partial T}{\partial x}\right)_t + \left(\frac{\partial \Theta}{\partial M}\right)_{t,T}\left(\frac{\partial M}{\partial x}\right)_t\right]\left(\frac{\partial T}{\partial x}\right)_t + D_T\left(\frac{\partial^2 M}{\partial x^2}\right)_t$$

$$+\left[\left(\frac{\partial D_T}{\partial M}\right)_{t,T}\left(\frac{\partial M}{\partial x}\right)_t + \left(\frac{\partial D_T}{\partial T}\right)_{t,M}\left(\frac{\partial T}{\partial x}\right)_t\right]\left(\frac{\partial M}{\partial x}\right)_t. \quad (7.92)$$

(Gradient of chemical potential)

The differentiation of the variable Θ is then performed as the product of the four variables in the definition of Θ. It is clearly necessary to have sorption-isotherm data over a range of temperatures to be able to evaluate H, E_L, and $\partial M/\partial H$. In the evaluation of these quantities, the sorption isotherm data can be fit to the equations of the Hailwood-Horrobin theory as described in Chap. 6.8. When the parameters K_1, K_2, W, A, B, and C are known as a function of temperature, the numerical determination of the derivatives may be accomplished.

Probably the most important practical application of nonisothermal unsteady-state moisture movement is in the drying of wood. Kollmann and Côté (1968) illustrate moisture-content and temperature profiles determined during the drying of wood based upon earlier work by Krischer (1942). The moisture profile is approximately parabolic with the lowest values on the surface and the highest in the center while the temperature profile is the reverse of this with the lowest value in the center. Therefore it would be expected that the thermal gradient would partially counteract the effect of the moisture-content gradient during the drying process.

7.8 Heat Transfer Through Massive Walls

The heat transferred through walls is usually calculated using the steady-state Fourier's law as discussed in detail in Chap. 5. This is usually satisfactory for well-insulated walls of light construction. However, in walls of massive construction with a high thermal capacity (product of specific heat, density, and thickness (c ϱ L) such as solid wood log walls, the storage of heat within the wall becomes a significant factor in the determination of the heat load. In order to take this into account, the unsteady-state Eq. (7.6) must be used together with appropriate boundary conditions for the inside and outside surfaces. The details of the mathematical method are described by Holman (1976). For a one-dimensional system, heat balances may be made at the interior and exterior surfaces by equating the rate of heat flow through the wall with the rate of gain or loss due to the film coefficient (includes convection and radiation) at the wall surface, calculated by the method described in Chap. 5.7. At the interior surface,

$$-KA \left.\frac{\partial T}{\partial x}\right|_{\text{interior wall}} = h_i(T_{wi} - T_i). \tag{7.60}$$

At the exterior surface,

$$-KA \left.\frac{\partial T}{\partial x}\right|_{\text{exterior wall}} = h_o(T_{wo} - T_o), \tag{7.61}$$

where h_i = interior film coefficient, h_o = exterior film coefficient (see Chap. 5.7 for typical values), T_{wi}, T_{wo} = temperatures of interior and exterior wall surfaces, T_i, T_o = inside and outside temperatures.

Heat-transfer calculations to determine time-temperature profiles may be made using Eqs. (7.60) and (7.61) for the boundary conditions, in combination with Eq. (7.6) for the interior of the wall. The equations may be transformed to the finite-difference form as described by Holman. The method outlined above is an approximation which will result in a small error because the heat storage of the finite elements adjacent to the surfaces is neglected. A more exact procedure which takes this storage into account is also explained by Holman. The values of T_i and T_o may be inserted as continous functions of time. Habayeb (1980) used approximately sinusoidal functions with periods of 24 h in the computer simulation of a residential building.

Several computer simulations have been conducted to elucidate the performance of massive walls in comparison with those of light construction. Goodwin and Catani (1979) compared similar residential buildings with heavy walls weighing 57 lb/ft^2 and light walls weighing 7 lb/ft^2 with different insulation levels. The heavy walls all had similar thermal storage capacities which were approximately six times greater than those of the light walls. Mass was concentrated alternatively on the inside, in the middle, and on the outside. Comparing massive and light walls of the same R value, it was found that the annual heat load was reduced in the massive walls with the greatest effect in warmer climates. For example, the annual savings for the entire building were computed as approximately 24% in Tampa, Florida and 2.5% in Boston, Massachusetts. The reductions are greater than these when walls only are considered because most of the heat losses are through other surfaces. It was also concluded that savings are greater when the mass is concentrated on the inside surface and that the effect of adding wall mass decreases as the overall R value is increased. Hopkins et al. (1979) compared one-storey office buildings of light and heavy envelope design and found that the overall annual energy requirements were not greatly affected by thermal mass. The calculated energy saving was 2.9% for Minneapolis, Minnesota, and 3.1% for Birmingham, Alabama, and Kansas City, Missouri. Rudoy and Dougall (1979) compared the effect of massive and light walls on energy requirements for single-family residential buildings in five locations in the U.S.A.

It was found that cooling requirements with heavy masonry walls were increased 2% to 4%, while heating requirements were reduced by 18% in Tampa, Florida, and 1% in Chicago, Illinois.

Burch et al. (1983) conducted a field study in which six test buildings of identical size (20 ft square) were constructed in Gaithersburg, Maryland. Each building had four storm windows, an insulated door, a ceiling with R-34 fiberglass insulation, and an insulated concrete slab foundation. Two buildings had uninsulated walls, one of wood frame construction with R-3.6 and the other of masonry with R-4.6. The other four had R values from 10.3 to 12.4, consisting of: insulated wood frame, insulated masonry with the mass outside, insulated masonry with the mass inside, and 7-inch-thick wood log construction. The results indicated no significant reduction in heat load during winter attributable to massive walls. On the other hand, during the intermediate seasons and during the summer cooling season there was a large reduction in both heating and cooling energy requirements due to the massive walls, with the greatest effect in the masonry walls with the mass on the inside in agreement with the results of Goodwin and Catani.

Habayeb (1980) conducted a computer-simulation study to evaluate dynamic performance of massive walls using the finite-difference equations which include the heat storage in the elements adjacent to the surfaces as described by Holman (1976). An essentially sinusoidal exterior temperature variation was assumed with a period of 24 h. The resulting sinusoidal response in the interior of the simulated residential building underwent a phase lag and a reduction in amplitude, both of which increase with the thermal capacity of the wall. These effects are functions of both the thermal diffusivity of the wall material and the wall thickness and plots are presented of the calculated results on a cloudy day with no internal mass and no windows. In agreement with the other investigations it was found that superior performance resulted when the mass was located on the inside and the insulation outside of a composite wall. It was pointed out, however, that common construction practice is the reverse of this with bricks or concrete on the exterior surface.

It is of interest to determine the time lag and amplitude reduction which would be obtained from a solid wood log construction with walls 8 inches (20 cm) thick. Assuming $G=0.5$ at $M=12\%$, Eq. (5.5) gives a thermal conductivity of 3.5×10^{-4} cal/(cm °C s), Eq. (7.7) gives a specific heat of 0.37 cal/(g °C), and the density is 0.56 g/cm^3. Therefore the specific heat on a volume basis (c ϱ) = 0.21 cal/(cm^3 °C) and the thermal diffusivity $(D_{hT}) = 1.7 \times 10^{-3}$ cm^2/s (Table 7.2). According to Habayeb's results, a time lag of 12 h is predicted. This would appear to be ideal for summer cooling because the interior response to the minimum exterior temperature during the night will occur during the heat of the day. Therefore the interior will be cooled during the day and warmed at night, greatly reducing the energy requirement for summer cooling. If the exterior temperature fluctuates with an amplitude of 10 °F, Habayeb's simulation indicates that an interior amplitude of 0.75 °F would be expected. Wood is an exceptionally good material in this respect because it combines a relatively low thermal conductivity with a high thermal capacity. A comparison of wood, concrete, brick, and fiberglass may be made by reference to Table 7.2. From this it is clear that both brick and concrete have relatively high thermal conductivities and it would therefore be impractical to obtain reasonable resistance (R-8, for example) without excessive thickness, although they

Table 7.2. Thermal properties of common construction materials

Materials	K cal/(cm °C s)	c cal/(g °C)	ϱ g/cm^3	$c\varrho$ cal/(cm^3 °C)	D_h cm^2/s	Thermal capacity for R-8 $c\varrho L$ cal/(cm^2 °C)
Wood $G=0.5\ M=12\%$	3.51×10^{-4}	0.37	0.56	0.21	1.7×10^{-3}	4.2
Concrete	22.3×10^{-4}	0.2	2.3	0.46	4.9×10^{-3}	
Brick	17.2×10^{-4}	0.19	1.9	0.36	4.8×10^{-3}	
Fiberglass	0.93×10^{-4}	0.23	0.032	0.0074	12.6×10^{-3}	0.041

both have very high thermal capacities. Therefore these materials are best used with a good insulator such as styrafoam, which is preferably placed on the exterior surface for energy economy during the intermediate seasons. Fiberglass is an excellent insulator with an R value of 3.9 per inch, but its thermal capacity is approximately 1% of that of wood. Therefore a time lag much less than the 12 h for wood and a much lower amplitude reduction would result from its use. Under these conditions the heating and cooling requirements may be calculated from the steady-state equations.

References

Adler E 1977 Lignin chemistry – past, present, and future. Wood Sci Technol 11:169–218
Adzumi H 1937 On the flow of gases through a porous wall. Bull Chem Soc Jpn 12:304–312
Ahlgren PA, Wood JR, Goring DAI 1972 The fiber saturation point of various morphological subdivisions of Douglas-fir and aspen wood. Wood Sci Technol 6:81–84
Ashrae 1977 ASHRAE handbook. Fundamentals. Am Soc Heating Refrigerating and Air-Conditioning Eng Inc New York, Ch 2, 22
ASTM 1968 ASTM standards; Standard methods of test for: Water vapor transmission of materials in sheet form E 96-66 Part 30; Water vapor transmission of thick materials C355-64 Part 14. Am Soc for Testing and Materials Philadelphia
Babbitt JD 1940 Observations on the permeability of hygroscopic materials to water vapor I. Observations at relative humidities less than 75%. Can J Res 18A(6):105–121
Bailey PJ, Preston RD 1969 Some aspects of softwood permeability I. Structural studies with Douglas-fir sapwood and heartwood. Holzforschung 23(4):113–120
Baines EF, Levy JF 1979 Movement of water through wood. J Inst Wood Sci 8(3):109–113
Banks WB 1970 Some factors affecting the permeability of Scots pine and Norway spruce. J Inst Wood Sci 5(1):10–17
Barber NF 1968 A theoretical model of shrinking wood. Holzforschung 22:99–103
Barber NF, Meylan BA 1964 The anisotropic shrinkage of wood. A theoretical model. Holzforschung 18:146–156
Bauer WD, Talmadge KW, Keegstra K, Albersheim P 1973 The structure of plant cell walls II. The hemicellulose of the walls of suspension cultured sycamore cells. Plant Physiol 51:174–187
Beal FC 1972 Density of hemicelluloses and their relationship to wood substance density. Wood Fiber 2:114–116
Behr EA, Sachs IB, Kukachka BF, Blew JO 1969 Microscopic examination of pressure-treated wood. For Prod J 19(8):31–40
Beiser W 1933 Mikrophotographische Quellungsuntersuchungen von Fichten- und Buchenholz an Mikromschnitten im durchfallenden Licht und an Holzklötzchen im auffallenden Licht. Kolloid Z 65:203–211
Blackwell J 1982 The macromolecular organization of cellulose and chitin. In: Brown RM Jr Cellulose and other natural polymer systems. Biogenesis, structure, and degradation. Plenum New York London, 403–428
Bolton AJ, Petty JA 1975 Structural components influencing the permeability of ponded and unponded Sitka spruce. J Microscopy 104:33–46
Bolton AJ, Petty JA 1977a Influence of critical point and solvent exchange drying on the gas permeability of conifer sapwood. Wood Sci 9:187–193
Bolton AJ, Petty JA 1977b Variation of susceptibility to aspiration of bordered pits in conifer wood. J Exp Bot 28:935–941
Bolton AJ, Petty JA 1978 A model describing axial flow of liquids through conifer wood. Wood Sci Technol 12:37–48
Bolton AJ, Petty JA 1980 A note on a possible source of error in maximum effective pore radius determination in conifer wood. Wood Sci Technol 14:45–47
Bonner LD, Thomas RJ 1972 The ultrastructure of intercellular passageways in vessels of yellow poplar (Liriodendron tulipifera) Part I: Vessel pitting. Wood Sci Technol 6:196–203
Bouveng HO, Meier H 1959 Studies on a galactan from Norwegian spruce compression wood (Picea abies Karst.). Acta Chem Scand 13:1884–1889
Bramhall G 1971 The validity of Darcy's law in the axial penetration of wood. Wood Sci Technol 5:121–134
Bramhall G, Wilson JW 1971 Axial gas permeability of Douglas-fir microsections dried by various techniques. Wood Sci 3:223–230

Brown JH, Davidson RW, Skaar C 1963 Mechanism of electrical conduction in wood. For Prod J 13:455–459
Brown RM Jr (ed) 1982 Cellulose and other natural polymer systems. Biogenesis, structure, and degradation. Plenum New York London, 519 pp
Burch DM, Remmert WE, Krintz DF, Barnes CS 1983 A field study of the effect of wall mass on the heating and cooling loads of residential buildings. Proc for the Thermal Mass Effects in Buildings Seminar, Knoxville 1982 Oakridge Nat Lab Oakridge. In press
Butterfield BG, Meylan BA 1982 Cell wall hydrolysis in the tracheary elements of the secondary xylem. In: Baas P (ed) New perspectives in wood anatomy. Martinus Nijhoff/W Junk, The Hague Boston London 71–84
Carpenter JH 1982 Fundamentals of psychrometrics. Carrier Corp, Syracuse 57 pp
Carrier WH 1911 Rational psychrometric formulae. Trans Am Soc Mech Eng 33:1005
Castellan GW 1966 Physical chemistry. Addison-Wesley Reading pp 120, 212
Cave ID 1972a A theory of the shrinkage of wood. Wood Sci Technol 6:284–292
Cave ID 1972b The influence of microfibril angle on the longitudinal shrinkage-moisture content relationship. Wood Sci Technol 6:293–301
Cave ID 1976 Modelling the structure of the softwood cell wall for computation of mechanical properties. Wood Sci Technol 10:19–28
Cave ID 1978a Modelling moisture-related mechanical properties of wood. Part I: Properties of the wood constituents. Wood Sci Technol 12:75–86
Cave ID 1978b Modelling moisture-related mechanical properties of wood. Part II: Computation of properties of a model of wood and comparison with experimental data. Wood Sci Technol 12:127–139
Choong ET 1962 Movement of moisture through softwood in the hygroscopic range. PhD Thesis SUNY Coll For Syracuse, 224 pp
Choong ET 1963 Movement of moisture through a softwood in the hygroscopic range. For Prod J 13:489–498
Choong ET 1965 Diffusion coefficients of softwoods by steady-state and theoretical methods. For Prod J 15:21–27
Choong ET, Skaar C 1969 Separating internal and external resistance to moisture removal in wood drying. Wood Sci 1:200–202
Choong ET, Skaar C 1972 Diffusivity and surface emissivity in wood drying. Wood Fiber 4:80–86
Comstock GL 1963 Moisture diffusion coefficients in wood as calculated from adsorption desorption, and steady-state data. For Prod J 13:97–103
Comstock GL 1965 Longitudinal permeability of green eastern hemlock. For Prod J 15:441–449
Comstock GL 1967 Longitudinal permeability of wood to gases and nonswelling liquids. F Prod J 17(10):41–46
Comstock GL 1968 Physical and structural aspects of the longitudinal permeability of wood. PhD Thesis SUNY Coll For Syracuse, 285 pp
Comstock GL 1970 Directional permeability of softwoods. Wood Fiber 1:283–289
Comstock GL, Côté WA Jr 1968 Factors affecting permeability and pit aspiration in coniferous softwood. Wood Sci Technol 2:279–291
Côté WA Jr 1963 Structural factors affecting the permeability of wood. J Polym Sci C-2:231–242
Côté WA Jr, Day AC 1969 Wood ultrastructure of the southern yellow pines. Tech Pub 95 SUNY Coll For, Syracuse 70 pp
Crank J 1956 Mathematics of diffusion. Clarendon Oxford, 347 pp
Cronshaw J 1960 The fine structure of the pits of Eucalyptus regnans (F. Muell.) and their relation to the movement of liquids into the wood. Aust J Bot 8:53–57
Davidson RW 1958 The effect of temperature on the electrical resistance of wood. For Prod J 8:160–164
de Groot SR 1962 Non-equilibrium thermodynamics. North-Holland Amsterdam, 510 pp
de Yong J 1982 Relative humidity and its measurement. Appita 35:483–490
Dunleavy JA, Mc Quire AJ 1970 The effect of water storage on the cell-structure of Sitka spruce (Picea sitchensis) with reference to its permeability and preservation. J Inst Wood Sci 5(2):20–28
Dunleavy JA, Moroney JP, Rossell SE 1973(7) The association of bacteria with the increased permeability of water-stored spruce wood. Br Wood Preserv Assoc Ann Conv 1–21
Dushman S, Lafferty JM (ed) 1962 Scientific foundations of vacuum technique. Wiley New York, 806 pp

Dye JL, Nicely VA 1971 A general purpose curvefitting program for class and research use. J Chem Educ 48:443-448

Ellwood EL, Ecklund BA, Zavarin E 1959 Collapse in wood and exploratory experiments to prevent its occurrence. Univ California. For Prod Lab Report

Erickson M, Larson S, Miksche G 1973 Zur Struktur des Lignins des Druckholzes von Pinus mugo. Acta Chem Scand 27:1673-1678

Eriksson O, Goring DAI, Lindgren BO 1980 Structural studies on the chemical bonds between lignins and carbohydrates in spruce wood. Wood Sci Technol 14:267-279

Erk S 1929 Über Zähigkeitsmessungen nach der Kapillarmethode. Z Tech Phys 10:452-457

Feist WC, Tarkow H 1967 A new procedure for measuring fiber saturation points. For Prod J 17(10):65-68

Freudenberg K 1968 The constitution and biosynthesis of lignin. In: Freudenberg K, Neish AC Constitution and biosynthesis of lignin. Springer, New York, 45-122

Freudenberg K, Neish AC 1968 Constitution and biosynthesis of lignin. Springer, New York, 129 pp

Gardner KH, Blackwell J 1974 The structure of native cellulose. Biopolymers 13:1975-2001

Gonzalez GE, Siau JF 1978 Longitudinal liquid permeability of American beech and eucalyptus. Wood Sci 11:105-110

Goodwin SE, Catani MJ 1979 The effect of mass on heating and cooling loads and on insulation requirements of buildings in different climates. ASHRAE Trans 85:869-884

Goring DAI 1981 Some aspects of the topochemistry of lignin in softwood and hardwoods. The Ekman Days 1981 (Stockholm), SPCI Rep 38, Vol 1:3-10

Goring DAI, Timell TA 1962 Molecular weight of native celluloses. Tappi 40:454-460

Griffin DM 1977 Water potential and wood-decay fungi. Ann Rev Phytopathol 15:319-329

Habayeb NA 1980 Influence of building materials on the thermal response of walls in passive solar cooled buildings. MS Thesis Syracuse Univ Syracuse, 58 pp

Harada H 1962 Electron microscopy of ultrathin sections of beech wood (Fagus crenata Blume). J Wood Res Soc 8:252-258

Harris JM, Meylan BA 1965 The influence of microfibril angle on longitudinal and tangential shrinkage in Pinus radiata. Holzforschung 19:144-153

Hart CA 1964 Theoretical effect of gross anatomy upon conductivity of wood. For Prod J 14:25-32

Hart CA, Thomas RJ Mechanism of bordered pit aspiration as caused by capillarity. For Prod J 17(11):61-67

Hawley LF 1931 Wood-liquid relations. US Dep Agr Tech Bull 284, 34 pp

Hendricks LT 1962 Thermal conductivity of wood as a function of temperature and moisture content. MS Thesis SUNY Coll For Syracuse, 83 pp

Henry WT 1973 Treating processes and equipment. In: Nicholas DD (ed) Wood deterioration and its prevention by preservative treatments II. Preservatives and preservative systems. Syracuse Univ Press Syracuse, 279-298

Hoadley RB 1967 Weather, water, and wood. Univ Mass Coop Ext Serv Pub No 15

Hoffmann P, Parameswaran N 1976 On the ultrastructural localization of hemicelluloses within delignified tracheids of spruce. Holzforschung 30:62-70

Hoffmann GC, Timell TE 1970 Isolation of a β-1,3-glucan (Laricinan) from compression wood of Larix laricina. Wood Sci Technol 4:159-162

Holman JP 1976 Heat transfer, 4th edn. McGraw-Hill New York, 530 pp

Hopkins V, Gross G, Ellifritt D 1979 Comparing the thermal performances of buildings of high and low masses. ASHRAE Trans 85:885-902

Howard ET, Manwiller FG 1969 Anatomical characteristics of southern pine stemwood. Wood Sci 2:77-86

Hudson MS, Shelton SV 1969 Longitudinal flow of liquids in southern pine poles. For Prod J 19(5):25-32

Jiang K-S, Timell TE 1972 Polysaccharides in compression wood of tamarack (Larix laricina) 4. Constitution of an acidic galactan. Svensk Papperstidn 75:592-594

Johansson MH, Samuelson O 1977 Reducing end groups in birch xylan and their alkaline degradation. Wood Sci Technol 11:251-263

Jost W 1960 Diffusion in solids, liquids, and gases. Academic Press New York, 558 pp

Katchalsky A, Curran PF 1965 Nonequilibrium thermodynamics in biophysics. Harvard Univ Press, Cambridge, 248 pp

Kaumann WG 1958 The influence of drying stresses and anisotropy on collapse in Eucalyptus regnans. Dev For Prod Tech Pap 3 CSIRO, Melbourne, 16 pp

Kellogg RM, Sastry CBR, Wellwood RW 1975 Relationships between cell-wall composition and cell-wall density. Wood Fiber 7:170–177

Kellogg RM, Wangaard FF 1969 Variation in the cell-wall density of wood. Wood Fiber 1:180–204

Kelso WC, Gertjejansen RO, Hossfeld RL 1963 The effect of air blockage upon the permeability of wood to liquids. Tech Bull 242 Univ Minnesota Agr Exp Stn, 40 pp

Kerr AJ, Goring DAI 1975 The ultrastructural arrangement of the wood cell wall. Cellul Chem Technol 9:563–573

Kininmonth JA 1971 Permeability and fine structure of certain hardwoods and effects on drying I Transverse permeability of wood to micro-filtered water. Holzforschung 25:127–133

Kollmann FFP, Côté WA Jr 1968 Principles of wood science and technology. I Solid wood. Springer, Berlin Heidelberg New York, p 231

Kollmann F, Malmquist L 1956 Untersuchungen über das Strahlungsverhalten trockener Hölzer. Holz Roh-Werkst 14:201–204

Krischer O 1942 Der Wärme- und Stoffaustausch im Trocknungsgut. VDI-Forschungsheft 415 Berlin

Kumar S 1981 Some aspects of fluid flow through wood 1 Mechanism of flow. Holzforsch Holzvert 33(2):28–33

Kuo C-M, Timell TE 1969 Isolation and characterization of a galactan from tension wood of American beech (Fagus grandifolia Ehrl). Svensk Papperstidn 72:703–716

Kutscha NP, Schwarzmann JM 1975 The lignification sequence of normal wood in balsam fir. Holzforschung 29:79–84

Lantican DM, Côté WA Jr, Skaar C 1965 Effect of ozone treatment on the hygroscopicity, permeability, and ultrastructure of western red cedar. Ind Eng Chem Prod Res Dev 4:66–70

Leyton L 1975 Fluid behaviour in biological systems. Clarendon Oxford, 235 pp

Liang CY, Bassett KH, McGinnes EA, Marchessault RH 1960 Infrared spectra of crystalline polysaccharides VII. Thin wood sections. Tappi 43:1017–1024

Liese W, Bauch J 1967a On anatomical causes of the refractory behavior of spruce and Douglas-fir. J Inst Wood Sci 19:3–14

Liese W, Bauch J 1967b The effect of drying on the longitudinal permeability of sapwood of gymnosperms. Souvenir Indian Plywood Res Assoc 35–40

Lin RT 1965 A study on the electrical conduction in wood. For Prod J 15:506–514

MacLean JD 1941 Thermal conductivity of wood. Heating, piping, and air conditioning 13:380–391

MacLean JD 1952 Preservative treatment of wood by pressure methods. US Dep Agr Handbook 40, 160 pp

Maku T 1954 Studies on the heat conduction in wood. Wood Res Bull 13, Wood Res Inst, Kyoto Univ

Marx-Figini M, Schultz GV 1966 Über die Kinetik und den Mechanismus der Biosyntheses der Cellulose in den höheren Pflanzen (nach Versuchen an den Samenhaaren der Baumwolle). Biochim Biophys Acta 112:81–101

Meier H 1962a Chemical and morphological aspects of the fine structure of wood. Pure Appl Chem 5:37–52

Meier H 1962b Studies on a galactan from tension wood of beech (Fagus silvatica L.). Acta Chem Scand 16:2275–2283

Meylan BA 1968 Cause of high longitudinal shrinkage in wood. For Prod J 18(4):75–78

Meylan BA 1972 The influence of microfibril angle on the longitudinal shrinkage-moisture content relationship. Wood Sci Technol 6:293–301

Mickelson RS 1964 Laminar, transition, and turbulent flow in capillary tubes. PhD Thesis. Wayne State Univ, Detroit (not seen)

Murmanis L, Chudnoff M 1979 Lateral flow in beech and birch as revealed by the electron microscope. Wood Sci Technol 13:79–87

Muskat M 1946 The flow of homogeneous fluids through porous media. Edwards, Ann Arbor, 763 pp

Nicholas DD, Siau JF 1973 Factors influencing the treatability of wood. In: Nicholas DD (ed) Wood deterioration and its prevention by preservative treatments. II Preservatives and preservative systems, Syracuse Univ Press Syracuse, 299–343

Nicholas DD, Thomas RJ 1968a The influence of enzymes on the structure and permeability of loblolly pine. Proc Am Wood Preserv Assoc 64:70–76

Nicholas DD, Thomas RJ 1968b Influence of steaming on ultrastructure of bordered pit membrane in loblolly pine. For Prod J 18(1):57–59

Norberg PH, Meier H 1966 Physical and chemical properties of the gelatinous layer in tension wood fibres of aspen (Populus tremula L). Holzforschung 20:174–178
Page DH, El-Hosseiny F, Bidmade ML, Binet R 1976 Birefringence and the chemical composition of wood pulp fibers. Appl Polym Symp 28:923–929
Palin MA, Petty JA 1981 Permeability to water of the cell wall material of spruce heartwood. Wood Sci Technol 15:161–169
Panshin AJ, de Zeeuw C 1980 Textbook of wood technology 4th edn. Mc Graw-Hill New York, 722 pp
Perilä O 1961 The chemical composition of carbohydrates of wood cells. J Polym Sci 51:19–26
Perng WR 1980a Studies on flow in wood I. Permeability and axial structural variation. Mokuzai Gakkaishi 26:132–138
Perng WR 1980b Studies on flow in wood. II Permeability and axial structural variation of short sample. Mokuzai Gakkaishi 26:219–226
Petty JA 1970 Permeability and structure of the wood of Sitka spruce. Proc R Soc Lond B 175:149–166
Petty JA 1973 Diffusion of non-swelling gases through dry conifer wood. Wood Sci Technol 7:297–307
Petty JA 1975 Relation between immersion time and absorption of petroleum distillate in a vacuum-pressure process. Holzforschung 29:113–118
Petty JA 1978a Effects of solvent-exchange drying and filtration on the absorption of petroleoum distillate by spruce wood. Holzforschung 32:52–55
Petty JA 1978b Influence of viscosity and pressure on the radial absorption of non-swelling liquids by pine sapwood. Holzforschung 32:134–137
Petty JA, Preston RD 1969 The dimensions and number of pit membrane pores in conifer wood. Proc R Soc Lond B 172:137–151
Phillips EWJ 1933 Movement of the pit membrane in coniferous woods, with special reference to preservative treatment. For 7:109–120
Prak AL 1970 Unsteady-state gas permeability of wood. Wood Sci Technol 4:50–69
Rasmussen EF 1961 Dry kiln operators handbook. US Dept Agric, Agric Handbook 188, 197pp
Resch H 1967 Unsteady-state flow of compressible fluids through wood. For Prod J 17(3):48–53
Robertson AA 1965 Investigation of the cellulose water relationship by the pressure plate method. Tappi 48:568–573
Rosen HN 1978 The influence of external resistance on moisture adsorption rates in wood. Wood Fiber 10:218–228
Rudman P 1965 Studies in wood preservation. I The penetration of liquids into eucalypt sapwoods. Holzforschung 19:5–13
Rudman P 1966 Studies in wood preservation. III The penetration of the fine structure of wood by inorganic solutions, including wood preservatives. Holzforschung 20:60–67
Rudoy W, Dougall RS 1979 Effects of the thermal mass on heating and cooling load in residences. ASHRAE Trans 85:903–917
Saiki H 1970 (Proportion of component layers in tracheid wall of earlywood and latewood of some conifers.) Mokuzai Gakkaishi 16:244–249
Saka S, Thomas RJ 1982 A study of lignification of loblolly pine tracheids by the SEM-EDXA technique. Wood Sci Technol 16:167–179
Sakakibara A 1977 Degradation products of proolignin and the structure of lignin. In: Loewus FA, Runeckles VC (eds) The structure, biosynthesis, and degradation of wood. Plenum Press, New York London, 117–139
Sarkanen KV, Ludwig CH (eds) 1971 Lignins. Occurrence, formation, structure, and reactions. Wiley-Interscience, New York, 916 pp
Sarko A, Muggli R 1974 Packing analysis of carbohydrates and polysaccharides. IV Valonia cellulose and Cellulose II. Macromolecules 7:486–494
Scheidegger AE 1974 The physics of flow through porous media 3rd ed. Univ Toronto Press, Toronto, 353 pp
Schneider A, Wagner L 1974 Bestimmung der Porengrößenverteilung in Holz mit dem Quecksilber-Porosimeter. Holz Roh-Werkst 32:216–224
Schreuder HA, Côté WA Jr, Timell TE 1966 Studies on compression wood. III Isolation and characterization of a galactan from compression wood of red spruce. Svensk Papperstidn 69:641–657
Sebastian LP, Côté WA Jr, Skaar C 1965 Relationship of gas phase permeability to ultrastructure of white spruce wood. For Prod J 15:394–404
Sebastian LP, Siau JF, Skaar C 1973 Unsteady-state axial flow of gas in wood. Wood Sci 6:167–173
Siau JF 1970a Pressure impregnation of refractory woods. Wood Sci 3:1–7

Siau JF 1970b A geometrical model for thermal conductivity. Wood Fiber 1:302–307
Siau JF 1971 Flow in wood. Syracuse Univ Press, Syracuse, 131 pp
Siau JF 1972 The effects of specimen length and impregnation time upon the retention of oils in wood. Wood Sci 4:163–170
Siau JF 1976 A model for unsteady-state gas flow in the longitudinal direction in wood. Wood Sci Technol 10:149–155
Siau JF 1980 Nonisothermal moisture movement in wood. Wood Sci 13:11–13
Siau JF 1983a Chemical potential as a driving force for nonisothermal moisture movement in wood. Wood Sci Technol 17:101–105
Siau JF 1983b A proposed theory of nonisothermal unsteady-state transport of moisture in wood. Wood Sci Technol 17:75–77
Siau JF, Babiak M 1983 Experiments on nonisothermal moisture movement in wood. Wood Fiber Sci 15:40–46
Siau JF, Davidson RW, Meyer JA, Skaar C 1968 A geometrical model for wood-polymer composites. Wood Sci 1:116–128
Siau JF, Kanagawa Y, Petty JA 1981 The use of permeability and capillary theory to characterize the structure of wood and membrane filters. Wood Fiber 13:2–12
Siau JF, Petty JA 1979 Corrections for capillaries used in permeability measurements in wood. Wood Sci Technol 13:179–185
Simpson WT 1973 Predicting equilibrium moisture content of wood by mathematical models. Wood Fiber 5:41–49
Simpson WT, Rosen HN 1981 Equilibrium moisture content of wood at high temperatures. Wood Fiber 13:150–158
Simson BW, Côté WA Jr, Timell TE 1968 Studies on larch arabinogalactan IV Molecular properties. Svensk Papperstidn 71:699–710
Simson BW, Timell TE 1978a Polysaccharides in cambial tissues of Populus tremuloides and Tilia americana II. Isolation and structure of oxyglucan. Cellul Chem Technol 12:51–62
Simson BW, Timell TE 1978b Polysaccharides in cambial tissues of Populus tremuloides and Tilia americana V. Cellulose. Cellul Chem Technol 12:127–141
Sjöström E 1981 Wood chemistry. Fundamentals and applications. Academic Press, New York, 223 pp
Skaar C 1948 The dielectrical properties of wood at several radio frequencies. Tech Pub 69, SUNY Coll For, Syracuse, 36 pp
Skaar C 1954 Analysis of methods for determining the coefficient of moisture diffusion in wood. For Prod J 4:403–410
Skaar C 1972 Water in wood. Syracuse Univ Press, Syracuse, 218 pp
Skaar C, Babiak M 1982 A model for bound-water transport in wood. Wood Sci Technol 16:123–138
Skaar C, Siau JF 1981 Thermal diffusion of bound water in wood. Wood Sci Technol 15:105–121
Smith DN, Lee E 1958 The longitudinal permeability of some hardwoods and softwoods. Dep Sci Ind Res For Prod Res Special Rep 13, London, 13 pp
Smulski SJ 1980 Woody cell wall penetration by a water-borne alkyd resin. MS Thesis, SUNY Coll Environmental Sci and For, Syracuse, 80 pp
Spolek GA, Plumb OA 1981 Capillary pressure in softwoods. Wood Sci Technol 15:189–199
Stamm AJ 1959 Bound-water diffusion into wood in the fiber direction. For Prod J 9:27–32
Stamm AJ 1964 Wood and cellulose science. Ronald New York, 549 pp
Stamm AJ, Hansen LA 1937 The bonding force of cellulosic materials for water (from specific volume and thermal data). J Phys Chem 41:1007–1016
Stamm AJ, Nelson RM 1961 Comparison between measured and theoretical drying diffusion coefficients for southern pine. For Prod J 11:536–543
Stamm AJ, Seborg RM 1934 Absorption compression on cellulose and wood. I Density measurements in benzene. J Phys Chem 39:133–142
Stayton CL, Hart CA 1965 Determining pore-size distribution in softwoods with a mercury porosimeter. For Prod J 15:435–440
Stone JE, Scallan AM 1967 The effect of component removal upon the porous structure of the cell wall of wood II. Swelling in water and the fiber saturation point. Tappi 50:496–501
Tanaka F, Koshijima T, Okamura K 1981 Characterization of cellulose in compression and opposite woods of a Pinus densiflora tree grown under the influence of strong wind. Wood Sci Technol 15:265–273

Tarkow H, Stamm AJ 1960 Diffusion through air-filled capillaries in softwoods Part I: carbon dioxide. For Prod J 10:247–250

Thomas RJ 1975 The effect of polyphenol extraction on enzyme degradation of bordered pit tori. Wood Fiber 7:207–215

Thomas RJ, Kringstad KP 1971 The role of hydrogen bonding in pit aspiration. Holzforschung 25:143–149

Thomas RJ, Nicholas DD 1966 Pit membrane structure in loblolly pine as influenced by solvent exchange drying. For Prod J 16(3):53–56

Tiemann HD 1906 Effect of moisture upon the strength and stiffness of wood. US Dep Agric For Serv Bull 70, 144 pp

Timell TE 1964 Wood hemicellulose I. Adv Carbohyd Chem 19:247–302

Timell TE 1965 Wood hemicellulose II. Adv Carbohyd Chem 20:409–483

Timell TE 1969 The chemical composition of tension wood. Svensk Papperstidn 72:173–181

Timell TE 1981 Recent progress in the chemistry, ultrastructure, and formation of compression wood. The Ekman Days 1981 (Stockholm), SPCI Rep 38, Vol 1:99–147

Timell TE 1982 Recent progress in the chemistry and topochemistry of compression wood. Wood Sci Technol 16:83–122

Tomkins EE 1974 Flow measurement utilizing multiple, parallel capillaries. In: Dowdell RB (ed) Flow: its measurement and control in science and industry. Instrument Soc Am, Pittsburgh, 465–471

Unligil HH 1972 Penetrability and strength of white spruce after ponding. For Prod J 22(9):92–100

USDA 1955 Wood handbook. US Dep Agric, Agric Handbook 72, 528 pp

USDA 1974 Wood handbook, US Dep Agric, Agric Handbook 72, 384 pp

Voight H, Krischer O, Schauss H 1940 Die Feuchtigkeitsbewegung bei der Verdunstungstrocknung von Holz. Holz Roh-Werkst 3:305–321

Wangaard FF 1943 The effect of wood structure upon heat conductivity. Trans Am Soc Mech Eng 65:127–135

Wangaard FF, Granados LA 1967 The effect of extractives on water-vapor sorption by wood. Wood Sci Technol 1:253–277

Ward RJ, Côté WA Jr, Day AC 1964 The wood substrate-coating interface. Proc Paint Res Inst 30. Official Digest 36:1091–1098

Wardrop AB, Davies GW 1961 Morphological factors relating to the penetration of liquids into wood. Holzforschung 15:129–141

Weast RC (ed) 1981–1982 CRC handbook of chemistry and physics, 62nd edn. CRC Press, Boca Raton

Weatherwax RC, Tarkow H 1968 Importance of penetration and adsorption compression of the displacement fluid. For Prod J 18(7):44–46

Wheeler EA, Thomas RJ 1981 Ultrastructural characteristics of mature wood of southern red oak Quercus falcata Michx.) and white oak (Quercus alba L.). Wood Fiber 13:169–181

Wilfong JG 1966 Specific gravity of wood substance. For Prod J 16(1):55–61

Wood JR, Goring DAI 1971 The distribution of lignin in stem wood and branch wood of Douglas fir. Pulp Pap Mag Can 72:T95–T102

Zimmermann MH 1983 Xylem structure and the ascent of sap. Springer, Berlin Heidelberg New York

Symbols and Abbreviations

A	Cross-sectional area, parameter in Hailwood-Horrobin equation, Ångstrom unit, intercept conductance
AH	Absolute humidity
AH_{dp}	Absolute humidity at saturation at dew point
AH_0	Absolute humidity at saturation
a	Square root of porosity, diameter of lumen, area fraction
a_v	Area fraction of pores
B	Parameter in Hailwood-Horrobin equation, intercept conductance
b	Constant in Klinkenberg equation, exponential coefficient in Bramhall permeability equation
C	Celsius temperature, parameter in Hailwood-Horrobin equation
C_w	Celsius wet-bulb temperature
c	Specific heat, concentration
D	Diffusion coefficient
D_a	Water-vapor diffusion coefficient in bulk air
D_B	Bound-water diffusion coefficient
D_h	Thermal diffusivity
D_p	Diffusion coefficient for hydrodynamic flow
D_v	Water-vapor diffusion coefficient in air in lumens
d	Diameter
E	Fractional change in concentration, temperature, or pressure squared (dimensionless potential)
$\bar{E}$	Fractional change in average concentration, temperature, or pressure squared (dimensionless average potential)
E_b	Activation energy for diffusion of bound water in the cell wall
E_L	Differential heat of sorption
E_v	Molar heat of vaporization of bound water in the cell wall
EMC	Equilibrium moisture content
F	Force, Fahrenheit temperature
F_{er}	Fractional external resistance to water-vapor movement
F_g	Fractional filling or evacuation with a gas
F_{VL}	Fraction of voids filled with liquid
F_w	Fahrenheit wet-bulb temperature
FSP	Fiber saturation point
f	Fractional available void volume
G	Specific gravity of moist wood
G_f	Specific gravity of wood at fiber saturation point
G_0	Specific gravity of ovendry wood
G'	Specific gravity of cell wall
g	Conductance, gravitational acceleration, gram

H	Quantity of heat, relative humidity
h	Relative vapor pressure, axial length of annulus
in	Inch
J	Joule
K	Specific permeability, equilibrium constant, thermal conductivity, Kelvin temperature
K_a	Thermal conductivity of air in lumens
K_c	Clausing factor
K_g	Thermal conductivity of wood
K_M	Conductivity coefficient for moisture movement in wood
K'	Thermal conductivity of cell-wall substance
k	Permeability, correction for gas expansion
k_g	Superficial gas permeability
k_{Lp}	Intercept longitudinal permeability of pit-opening portion of wood specimen
k_{Lt}	Intercept longitudinal permeability of tracheid portion of wood specimen
k'_{Lp}	Longitudinal superficial gas permeability of pit-opening portion of wood specimen
k'_{Lt}	Longitudinal superficial gas permeability of tracheid portion of wood specimen
L	Length of specimen in flow direction
L_{Lp}	Portion of specimen length corresponding to flow through pit membranes
L_{Lt}	Portion of specimen length corresponding to flow through tracheid
L_p	Thickness of pit membrane
L_t	Length of tracheid
l	slope
M	Moisture content of wood as percent of ovendry weight
M_f	Fiber saturation point
M_{max}	Moisture content with voids completely filled
M_w	Molecular weight
m	Meter, coefficient for kinetic-energy losses, slope
mm	Millimeter
N	Number of vessels or capillaries
n	Number of vessels per cm^2 of cross section, number of moles
n_p	Number of conductive pit openings per cm^2 of cross section
n_{p_T}	Number of conductive pit openings per cm^2 for tangential flow
n_{pt}	Number of conductive pit openings per tracheid
n_t	Number of tracheids per cm^2 of cross section
n_{tc}	Number of conductive tracheids per cm^2 of cross section
nm	Nanometer
P	Pressure
P_a	Atmospheric pressure
P_0	Pressure on gaseous side of liquid-gas interface
P_1	Pressure on liquid side of liquid–gas interface
P	Dimensionless pressure
p	Partial vapor pressure
p_0	Saturated vapor pressure of water

p_{ow}	Saturated vapor pressure at wet-bulb temperature
Q	Volume rate of flow
R	Variable radius, thermal resistance (ft² h °F/BTU), radial shrinkage
Re	Reynolds' number
Re′	Critical Reynolds' number of turbulence
Re″	Critical Reynolds' number for nonlinear flow due to kinetic energy losses
R_{SI}	Thermal resistance (m²K/W)
R	Universal gas constant
r	radius, resistivity of wood
r_i	Radius of gas-liquid meniscus, inside radius of annulus
r_t	Radius of tracheid
r'	Resistivity of cell wall
S	Surface emission coefficient, percent swelling of wood
S'	Percent swelling of cell wall
s	Percent shrinkage of wood, slip-flow factor
s'	Percent shrinkage of cell wall
T	Temperature, tangential shrinkage
T_{dp}	Dew-point temperature
T_i	Initial temperature, inside temperature
T_0	Outside temperature
T_{wi}	Temperature of inside wall
T_{wo}	Temperature of outside wall
t	Time
U	Thermal conductance
V	Volume
v	Volume fraction, linear velocity, specific volume
v_a	Porosity
v_w	Volume fraction of cell-wall substance
W	Moisture content of air, parameter in Hailwood-Horrobin equation, watt, work
WB	Wet-bulb temperature
w	Mass
w_1	Fraction of cross wall which is effective for conduction
X	Dimensionless distance
x	Variable distance along the flow direction
y	Distance of separation
Z	Total fraction of cross wall effective for conduction
z	Height of capillary rise

Greek Letters

α	Fraction of tracheid overlap
γ	Surface tension
ε	Dielectric constant of wood
ε'	Dielectric constant of cell wall
η	Viscosity
θ	Contact angle
λ	Mean free path
μ	Chemical potential

μm	Micrometer (micron)
ϱ	Density
ϱ_L	Density of liquid
ϱ_w	Normal density of water
τ	Dimensionless time
ψ	Water potential

Subscripts

f	At fiber saturation point
L	In the fiber direction
o	Oven-dry
R	In the radial direction
T	In the transverse or tangential direction

Subject Index

Abies alba, pit aspiration   115
A. grandis, resistance of diffusion paths   162
Absolute humidity   4–6, 20
Activated moisture content   166
Activation energy for diffusion   155, 157, 166
Adsorption-desorption ratio   20, 153
Adzumi equation   84, 86, 89
Air blockage   95, 100, 111
Air velocity (see also Surface resistance)
   effect on diffusion   153, 161, 174, 186, 190
   effect on heat conduction   148
   effect on wet-bulb temperature   7
Ash content of wood   35
Aspiration, pit   115–118
   causes of   113–117
   contributing factors   118
   mechanism   116–118
   prevention   115, 116
   resistance to   41, 97, 98, 115
   reversal of   115, 116
   TEM   47, 50

Balsa wood, fiber saturation point   125
Barometric pressure, effect of altitude   10, 13
Betula sp., change in lumen size with M   31
B. papyrifera
   flow properties   61
   tension wood, composition   71, 72
Boiling of water   4, 5
Bound water (see also Diffusion; Moisture content, Unsteady-state equations)
   compaction at low moisture contents   26–29, 122
   specific gravity, graph   27
   concentration in cell wall   159
Boundary layers, turbulent and viscous   147
Bubble-point test   95, 118
Bulk flow, definition   73

Capillaries
   distribution of sizes in wood   108
   in cell wall   25–28, 52, 62
   reduced relative humidity at saturation   121–123
   rise and depression of liquids   107, 112
   size from flow measurements   49, 50, 86–91, 93–96
   velocity distribution   79, 80, 83

Capillary flow   76
   corrections for short capillaries   88, 89
   during drying of wood   131
   Knudsen diffusion   75, 78, 84–89
   nonlinear flow   74, 77, 78, 83–85
   turbulent flow   82, 83
   viscous flow   77–82
Capillary forces
   across pit membranes   111–118
   during drying   97–99, 111–118
   effect on collapse   111–114
   effect on liquid transport   100, 111
   effect on pit aspiration   115–118
   graphs
      mercury-air interface   110
      water-air interface   109
   related to capillary radius   106–111
Capillary tension (see Capillary forces)
Capillary water (see Free water)
Carrier's equation   10
Cell wall (see also S_1, S_2, S_3 layers)
   composition   35, 72
   diffusion coefficient   152–155
   dimensions   44
   penetration of polymers   125, 126
   permeability   62, 63, 102
   porosity   25–28
   shrinkage and swelling   30, 31
   specific gravity
      dry   25–29, 30, 31
      moist   31, 160
   structure   36–40
   thermal conductivity   135, 139, 144–146
   thickness from porosity   30
Cellulose   35–36
   cellulose I and II   64
   composition   63, 64
   crystallinity   64
   degree of polymerization   64
   density   28
   dimensions of unit cell   64
   structural formula   64, 65
Chamaecyparis lawsonia, mercury porosimetry   108
Chemical potential (see also Water potential)
   effect of relative humidity   167
   equations   119, 152
   gradient for nonisothermal diffusion   166, 167

Subject Index

Chemical potential
   temperature dependency   119, 167
Clausing factor   88, 89
Coefficients, diffusion (see Diffusion coefficients)
Collapse
   contributing factors   114
   mechanism   111–114
   reversal of   114
Compression wood   39, 70–72
Comstock model   91–93, 103, 131
Condensation (see also Vapor barriers)
   in buildings   24
   in capillaries   121–128
   in wood   174
Conductance, air films   148
Conductance and conductivity   136, 137
Conductivity, thermal (see Thermal conductivity)
Contact angle   107–109
Convection   147, 148
Couette correction   80, 84, 88, 89
Crystallites   64, 66

Dalton's law   2, 4
Darcy, unit of permeability   78
Darcy's law
   annulus   76, 183, 218, 219
   assumptions   74
   exceptions   78, 82–95, 100–103, 216, 217
   steady state
      gases   74–76
      liquids   74
   unsteady state
      gases   179–183
      liquids   213–219
Density (see also Specific gravity)
   cell wall   25–29, 31
   effect of moisture content   25
   relationship to specific gravity   25
Dew point
   definition   13
   from psychrometric charts   9–12
   in heated buildings   24
   relationship to relative humidity   6, 18
Dielectric constant of wood, equations   149, 150
Diffusion (see also Diffusion coefficients; Nonisothermal diffusion; Unsteady-state equations; Unsteady-state transport)
   above boiling point and FSP   162
   bound water   73, 152–155
   definition   151
   interdiffusion of gases   73, 157, 165
   Knudsen   75, 78, 84–89
   nonisothermal   159–165, 206–208
   resistance of tracheids and pits   161, 162
Diffusion coefficients
   directional ratios, L/T   207, 211
   gas flow   180, 181
   hindered in pits   162
   thermal diffusivity   176, 177, 224
   water-vapor
      activation energy   155, 157
      air in lumens   157–160
      alternate gradients   151
      average value, $\bar{D}$ 154, 155, 172
      bulk air   151–154
      cell wall   154, 155
      construction materials, table   173
      conversion to various gradients   151, 152, 159, 160
      directional ratios
         cell wall   155, 160
         wood   206–211
      effect of adsorption and desorption   153
      effect of fiber direction
         cell wall   155, 160
         wood   164, 165
      effect of moisture content   152–155, 157
      effect of specific gravity   160, 161, 163–165, 212
      effect of temperature   155, 158
      equations
         air in lumens   160
         from activation energy   155, 160
         longitudinal model   165
         transverse model   161
      equilibrium between lumen and cell wall   159
      measurement
         effect of adsorption and desorption   153
         steady-state method   153, 171–174
         unsteady-state method   152, 153, 186, 190
      relationship to permeance   173, 174
Diffusion equations (see Unsteady-state equations)
Dimensionless quantities
   distance   181
   Grashof number   147
   Nusselt number   147
   potential   184, 185
   Prandtl number   147
   pressure   182
   Reynolds' number   77, 78, 82–85, 91, 147
   time   181, 184, 196, 214, 218
Drying
   collapse during   111–114
   effect on permeability   97, 98, 111–118
   freeze-drying   98
   low-surface-tension   98, 99, 112, 115, 116
   removal of free water   112–114, 131
Drying oven   24

E and $\bar{E}$, definitions   184, 185
Earlywood and latewood   35–44

electromicrographs 38, 47, 55, 56
permeability 35, 52, 53, 61, 97, 98, 115–118, 202
pit pairs 41–43, 47, 49, 51
pressure to deflect pit membranes 97, 98, 116
tracheids 43
Electrical resistance of wood 149
End effects (see Surface resistance)
Enzymatic degradation of pit membranes 99
Equilibrium moisture content (see also Fiber saturation point) 1, 19–24, 124–131
  above 98% relative humidity 121, 125–128
  definition 20
  derivative relative to H 169
  effect of seasons and location 24
  effect of temperature 21
  effect of mechanical stress 126, 128
  graphs 21, 127, 130
  in saturated air 121, 124, 126, 127
  relationship to water potential 124–128
  sorption and desorption 20, 21
  tables, below and above boiling point 14, 15, 22, 23
  wood in soil 130
Eucalyptus sp., flow study 62
E. deglupta, zones of differing permeability 215, 216
E. regnans
  collapse 111, 114
  flow study 62
  permeability 104
Evaporation of water 1–5
Extractives 35
  effect on equilibirium moisture content 20
  effect on permeability 51, 62

Fagus crenata, pit membrane dimensions 60, 61
F. grandifolia, tension wood fiber, TEM 40
F. sylvatica
  permeability 104
  pit membrane opening sizes 59
  treatability 62
Fiber saturation point (see also Equilibrium moisture content)
  balsa wood 125
  decrease with temperature 156
  definition
    based on changes in properties 20, 128
    Tiemann's 122
  effect of extractives 20
  effect of specific gravity 20, 33
  effect on shrinkage and swelling 30–34
  measurement methods
    changes in physical properties 20, 124
    microscopic measurements 126
    nonsolvent water method 125, 126
    polymer-exclusion method 125

pressure-plate method 125, 126
  relationship to water potential 124–128
Fibers 54–58
  dimensions 57
  fiber tracheids 54
  libriform fibers 54, 58
  treatability 58, 103
Fick's laws of diffusion
  alternative potentials 151, 152, 174
  application to nonisothermal diffusion 117, 166
  exceptions 155–158, 166
  unsteady-state 177–179
Film coefficients, thermal 147–149, 224
Finite difference equations 178, 181–183, 223, 224
Fire performance of wood 133
Flow of fluids, types (see also Diffusion; Permeability)
  capillary flow during drying 112–114, 131
  Knudsen diffusion 75, 78, 84–89
  nonlinear 74, 77, 78, 83–85, 91, 102
  turbulent 77, 82, 83
  viscous 73–82
Fourier's law
  steady-state 132–134
  unsteady-state 175–178
Free water (see also Moisture content)
  definition 1, 20, 21
  measurement 125, 126
  removal 112–114, 131
Freeze drying 98

G-layer 40, 71
Galactan 70–72
Galactoglucomannan 63–66, 71
Glucomannan 66–68, 71
Gum canals 57

Hailwood-Horrobin theory 168, 169
Hardwoods
  cells, photomicrographs 58
  chemical composition 63, 67–69
  diffuse-porous 53–55
  flow studies 61–63, 103, 104
  pitting 58–61
  ring-porous 53–56, 103, 104
  structure 53–61
  TEM 55, 56
  tension wood 40, 71
  volumetric composition 57, 58
Heartwood and sapwood, permeability 35, 52, 53, 63, 98, 99
Heat load for thick wood walls
  computer simulations 223–225
  time lag for conduction 224
Heat of sorption 168

Heat of vaporization, bound water 2, 157, 168
Heat of wetting 124, 168
Heat transfer (see also Fourier's law; Thermal conductivity; Unsteady-state equations; Unsteady-state transport)
  steady-state 132–149
    equations
      empirical 134, 135, 143
      geometrical model 135–146
    R and U values 146–149
    table, thermal conductivity 133
  unsteady-state 175–177, 184, 206
    in wood 203–206
    massive walls 223–225
    thermal capacity 225
Hemicelluloses 35, 36, 65–68
  compression wood 70, 71
  density 28
  hardwood 67
  softwood 66–68, 71
  tension wood 71, 72
Humidity (see Absolute humidity; Relative humidity)
Humidity ratio (see also Moisture content, air) 4–12
Hydrogen bonding 20, 116–118, 152
Hygroscopic water (see Bound water; Moisture content)

Insulation for buildings 147
  table 133
  wood 133, 148, 223–225

Juglans nigra, collapse 114
Jurin's law 106, 107
Juvenile wood 72

Klinkenberg equation 87–90, 94, 181–183
Knudsen diffusion 75, 78, 84–89
  calculation of radius 85–88, 93, 94
  in liquids 85

Larix laricina
  compression wood tracheid, TEM 39
  pit pairs, TEM 46
L. occidentalis, SEM 42
Latewood (see Earlywood and latewood)
Lignin 35–37
  compression wood 71
  density 28
  distribution in wood 72
  hardwood 69, 70
  softwood 68, 69
  structural formulas 68–70
Liquidambar styraciflua, volumetric composition 57

Liriodendron tulipifera
  pit membrane size 59
  TEM 55
Log homes
  condensation 174
  heat load 223–225
  insulation, steady-state 148
  time lag for conduction 224
  reduction in temperature amplitude 224
Low surface-tension drying 98, 99, 112, 115, 116
Lumens (see also Fibers; Tracheids; Vessels)
  relative resistance to bulk flow 85
  relative resistance to diffusion 161, 162
  shrinkage and swelling 31

Macrofibril 36
Margo 49–51
Massive walls, heat transfer 223–225
Mean free path 78
  air at 20 °C 85
  effect on Knudsen diffusion 87
  liquids 85
Mechanical strength, effect of moisture content 124
Meniscus 107
Mercury porosimetry 26, 28, 108
Microfibrils 36, 37, 64, 66
Microvoids in cell wall (see also Capillaries; Cell wall) 26–28, 52, 53, 62
Middle lamella – primary wall 36, 37, 72
Models
  cell wall, interrupted lamellar 37
  dielectric constant 149, 150
  electrical resistivity 144
  permeability
    Bramhall 100–102, 203
    Comstock 91–94, 103
    parallel-uniform-circular capillary 81, 82, 85–89
    Petty 89–91, 93–94, 103, 175, 176
  thermal conductivity
    cellular longitudinal 143, 146
    cellular transverse 135–143
    parallel conductance 135
  water-vapor diffusion
    longitudinal 157–159
    transverse 160, 161
Moisture characteristic, graphs (also see Sorption isotherm)
  soil 129
  wood 127
Moisture content (see also Equilibrium moisture content; Fiber saturation point)
  air 4–12
  bound and free water 1, 20, 21
  definition 19

effect on bound-water diffusion 152–155, 158, 164
effect on cell-wall porosity 29
effect on diffusion through air in lumens 157–160
effect on diffusion through wood 164, 165
effect on heat of wetting 124
effect on lumen area 31
effect on porosity 28, 29
effect on shrinkage and swelling 30–34
effect on specific gravity and density 25, 31–33
effect on specific heat 177
effect on strength 124
effect on thermal conductivity
   cell wall 139
   wood 134, 135, 139, 143
wood 19–21
Moisture meters 149, 150
Moisture movement (see also Capillary flow; Diffusion; Diffusion coefficient; Free water; Nonisothermal diffusion; Permeability)
  above fiber saturation point 95, 112–114, 131, 162
  below fiber saturation point 151–165

Nonisothermal diffusion
  applications
    drying 171
    log homes 170, 171
  dT/dM vs. H, graphs 170
  equations
    gradient of activated moisture content 166, 167
    gradient of chemical potential 167, 168
  experimental results
    encapsulated wood specimens 165, 166
    fiberboard 165
    solid wood 167
    gaseous interdiffusion 165
    unsteady-state 221–223
Nonlinear flow 74, 77, 78, 83–85, 91, 102
Nothofagus fusca, permeability 104

Oaks (see also Quercus sp.)
  red, permeability 61, 81, 103, 104
  white, permeability 57, 104
  zones of different permeability 103
Osmotic pressure 102, 128
Oven-dry, definition 19
Ozone-oxygen treatment 99

Parenchyma cells 40, 58, 59, 61–63
Partial vapor pressure 1–4, 24
Pectin 65, 71
Perforation plates 53–57
Perm, definition 173

Permeability (see also Darcy's law; Poiseuille's law; Unsteady-state equations; Unsteady-state transport) 73–104
  annulus 76
  apparatus 96, 97
  cell wall 62, 63, 102
  correlation with treatability 103
  definition 73
  directional ratios
    longitudinal to radial 103, 104
    longitudinal to tangential 45, 92, 93, 103, 104
    radial to tangential 53, 103, 104
  earlywood and latewood 35, 52, 53, 61, 97, 98, 114, 116–118, 215, 216
  effect of drying 97, 98, 111–118
  effect of moisture content 100
  effect of specimen length 100–102
  fibers 58, 103
  flow studies
    hardwoods 61–63
    softwoods 52, 53
  gas 74–76, 85–96, 179–183
  hardwoods 81, 100, 103, 104
  liquid 74, 76, 95, 111, 213–220
  measurement 95–98
  rays 44, 52, 53, 61–63, 103, 104
  relationship to capillary radius 81, 82
  relationship to porosity 73, 82, 100
  relationship to size of pit openings 89–91, 93–95, 113
  sapwood and heartwood 52, 53, 61–63, 98, 99
  specific
    definition 78
    measured with gas and liquid 95
  treatments to increase 98–100
  units, conversion 78
  unsteady-state 179–183, 213–220
  vessels 54, 61–63, 82, 103
  water vapor 173, 174
  zones in wood 103, 181, 215, 216
Permeance 173, 174
  relationship to diffusion coefficient 173, 174
  table of values 173
Petty permeability model 89–91, 93, 103
  applied to unsteady-state gas flow 181, 182
Picea sp.
  change in lumen size with M 31
  permeability 52
P. abies
  permeability of cell wall 102
  pit aspiration 115
  radial permeability 44, 52
P. engelmannii, mercury porosimetry 108
P. glauca
  size of pit-membrane openings 95

P. glauca
  permeability to gas and nonswelling liquids 95
  effect of length on permeability 100
P. mariana, determination of fiber saturation point 125–127
P. sitchensis
  bound-water diffusion coefficient 152, 153
  characterization of structure from flow measurements 91
  effect of solvent-exchange drying on permeability 99
  fiber saturation point 20, 124, 125
  fraction of conducting tracheids 162
  ponding to increase permeability 99
  resistance of diffusion paths 162
  size of pit membrane openings 50
  Stamm's measurement of fiber saturation point from physical properties 124
Pinaceae family
  pit aspiration 97, 98
  presence of torus 49, 97
Pinus sp., treatability 52, 53
P. elliottii
  bordered pit pair, TEM 47
  structure of torus surface, TEM 51
P. eschinata, bordered pit pair, TEM 47
P. glabra, bordered pit pair, TEM 48
P. pungens, bordered pit pair, TEM 48
P. radiata, treatability 52
P. resinosa, solvent-exchange drying 116
P. rigida
  bordered pit pair, TEM 50
  half-bordered pit pair, TEM 46
P. strobus
  compression wood, composition 71
  nonisothermal moisture diffusion 167
  volumentric composition 44
P. sylvestris
  penetration vs. time 215, 216
  pit aspiration during drying 115
  pressure to deflect pit membranes 97, 98, 117
  radial permeability 44
  zones of different permeability 215, 216
P. taeda
  effect of length on permeability 100–102
  effect of steaming 99, 100
  factors influencing pit aspiration 115, 116
  mercury porosimetry 108
  structure 40, 41, 43
Pit membranes (see also Aspiration, pit; Pit pair)
  bacterial treatment 99
  capillary forces at openings 108, 114
  composition 72
  deaspiration by steaming 99
  enzyme treatment 99
  incrustations 99

occlusion 51, 72
ozone treatment 99
pressure to deflect 97, 98, 117
resistance to diffusion 161, 162
resistance to flow 49, 50, 85, 90, 91
size of openings
  hardwood 58–60
  softwood 49–51
thickness 50, 51, 61, 88, 91, 94, 95
torus, occurrence of 49, 97

Pit pair (see also Aspiration, pit; Pit membrane)
  definition 45
  dimensions 44, 46–51, 59–61
  earlywood and latewood 46, 50, 97, 98
  effect on diffusion 161, 162
  hardwood 58–61
  number of conducting pairs per tracheid 94
  number per tracheid 43
  resistance to diffusion 161, 162
  resistance to flow 49, 50, 85
  softwood 45–51
  TEM 46–51, 59, 60
  types 45, 46
Poise 78, 79
Poiseuille's law 79–82
Polar and nonpolar liquids 26–29, 52, 53, 59–62, 74, 102
Ponding 99
Porosity (see also Cell wall; Specific gravity)
  cell wall 25–28
  relationship to cell wall thickness 30
  relationship to dielectric constant 149, 150
  relationship to electrical resistance 149
  relationship to permeability 73, 82
  relationship to specific gravity, graphs 141, 142
  relationship to thermal conductivity 135, 136, 138, 143–146
  wood 28–30
Prescap® treatment 111
Pressure impregnation
  effect of capillary forces 111
  importance of rays 45, 52, 53, 61–63, 103, 104
  relationship to permeability 103
  retention 29, 30, 210–213
Prosenchyma cells 40, 54
Pseudotsuga menziesii
  coastal and interior 98, 99
  decrease in permeability with length 100, 102, 216, 217
  fiber saturation point 126
  flow study 52, 53
  permeability 98, 100, 101, 103, 104
  pit aspiration 98, 99, 114
  unsteady-state gas flow 181
Psychrometric chart 7–12, 18

Quercus falcata, bordered pit, TEM 59
Q. rubra
  intervessel pitting, TEM 60
  structure 56, 57
  treatability 61

R, values of 119, 152, 160
R-value 146–148
Radiation 147, 148
Rays (also see Hardwoods; Permeability; Softwoods; Structure)
  composition 72
  effect on shrinkage 45
  multiseriate 53–56
  permeability 44, 52, 53, 61–63, 103, 104
    hardwood 61–63, 104
    softwood 44, 52, 53, 103
  volume fraction in wood 43, 44, 58
Reaction wood (see Compression wood; Tension wood)
Red oak (see also Oaks; Quercus sp.)
  moisture characteristic 126, 127
  permeability 54, 81, 103, 104
Reduced vapor pressure in capillaries 121–123
Relative humidity (see also Equilibrium moisture content; Partial vapor pressure; Psychrometric chart; Saturated vapor pressure; Sorption isotherm; Water potential)
  control 18, 19
  definition 5, 6
  effect of changes in pressure and temperature 21, 24
  effect on wood moisture content 1, 14, 15, 19–24, 124–128
  from psychrometric charts 9–12
  in heated buildings 24, 25, 170, 171
  over capillary meniscus 121–123
  over solutions, tables 19
  relationship to absolute humidity in heated buildings 24
  seasonal changes 24
  wet and dry-bulb temperatures, tables 14–17, 22, 23
Resin canals 40, 41, 43, 51
Resistance (electrical) 124, 149
Resistance to flow, distribution between pits and lumens 85, 90, 91
Resistivity 136, 137
Retention of liquids 29, 30, 210–213
Reynolds' number
  critical values 77, 78, 82–85, 147
  definition
    circular capillary 77
    plane surface 147, 148

S_1, S_2, S_3 layers 35–39, 44
  distribution of polysaccharides 71, 72

Sap flow 108, 119, 128, 131
Sapwood and heartwood permeability 52, 53, 61–63, 98, 99
Saturated vapor pressure 1–5, 24
  equation 2
  tables 3, 4
Sebastian equation for transient gas flow 181, 182
Sequoia sempervirens
  collapse 112, 114
  extractives 35
  fiber saturation point 20
  radial permeability 52
Shrinkage and swelling
  cell wall 30, 31
  directional ratio 33, 34, 37, 45
  longitudinal 33, 37–40
  lumens 31
  microfibrillar strands in margo 100
  wood 24, 31–34, 37–39, 62
Sling psychrometer 7–13
Slip-flow 75, 78, 84–89, 93–95
Softwoods
  cells 41–44, 58
  chemical composition 63, 65–69
  compression wood 70–72
  flow studies 51–53
  pitting 45–51
  SEM 42
  structure 40–45
  volumetric composition 44
Soil
  moisture characteristic, graph 129
  sorption isotherm with wood, graph 130
Solvent-exchange drying 27, 28, 98, 99, 115, 116
Sorption isotherm (see also Equilibrium moisture content; Sorption isotherm)
  adsorption and desorption 20, 21
  caculation of derivative 168, 169
  definition 20
  effect of temperature 21, 156
  graphs
    wood in air 21, 127
    wood in soil 130
  relationship to chemical potential 167, 170, 222
  table with wet and dry bulb data 14, 15, 22, 23
Specific gravity
  bound water 27–29
  cell wall
    dry 25–28, 30, 31
    moist 160
  definition 25
  effect on diffusion 160, 161, 163–165, 212
  effect of moisture content 25, 26, 33, 160

Specific gravity
  effect on thermal conductivity   134–136, 144–146
  influence of shrinkage and swelling   31–34
  relationship to density   25
  relationship to porosity   28–30
    graph, G vs. $\sqrt{v_a}$ 141
    graph, G vs. v_a 142
Specific heat   177, 225, 226
Specific permeability (see Permeability)
Stamm's measurements of bound-water diffusion   152–155, 158
Steady-state transport (see Diffusion; Diffusion coefficient; Permeability; Thermal Conductivity)
Steaming   99, 100
Structure (see also Cell wall; Hardwoods; Pit pairs; Softwoods)
  cell wall   36–40
  hardwoods   53–59
  pitting
    hardwood   58–61
    softwood   40–51
Surface resistance (see also Boundary layers; Clausing factor; Couette correction; Film coefficients)
  bulk flow   80, 84, 88, 89
  diffusion of moisture, effect of air velocity and thickness   153, 161, 174, 186, 190
  heat transfer   147, 148, 224
Surface tension (see also Capillary forces; Freeze drying; Solvent-exchange drying)
  definition   105, 106
  mercury-air   107–109, 110, 112
  sap   118
  table   112
  water-air   108, 109, 112
Swelling (see Shrinkage and swelling)
Symbols and abbreviations   233–236

Temperature
  dew-point   6–14, 18, 24
  saturated steam   3, 4
  wet and dry bulb   7–18, 22, 23
Tension, capillary (see Capillary forces)
Tension wood
  composition   71, 72
  TEM   40
Thermal capacity, table   210
Thermal conductivity (see also Fourier's law; Heat transfer)   132–149
  air in lumens   135, 139
  air space, large   148, 149
  bound and free water   134, 135
  cell wall   135, 139, 144–146
  directional ratio, L/T   134, 146

graphs
  K_{gL}/K_{gT} vs. v_w, model equations   146
  $K_{gT}/$ vs. a, model equation and linear function   145
  K_{gT} vs. v_w, model equation   144
  K_{gT} vs., v_w, values from literature   143
heat loss from buildings   148, 149
MacLean's equations   134, 135
model equations   135–146
  longitudinal   143, 146
  transverse   137–143
nonuniformity of flux in cell wall   138, 139
relationship to porosity   135, 136, 138, 143–146
table   133
units and conversions   132
wood
  as an insulator   146, 147
  values from literature   143
Thuja occidentalis, lack of pit aspiration   115
T. plicata
  collapse   114
  fiber saturation point   20
  ozone-oxygen treatment   99
Tilia americana
  fiber saturation point   20
  permeability   81, 103, 104
  pit-membrane opening size   59
Topochemistry of wood   72
Torus (see also Aspiration, pit; Pit membranes)
  chemical composition   72
  diameter   47–49
  dishing   50, 117
  occurrence   49, 97
  openings   49, 117
  thickness   47, 48
Tracheids (see also Softwoods)
  compression wood, TEM   39
  dimensions   44
  earlywood and latewood   38, 41–43
  fraction conducting   94, 162
  length to diameter ratio   93
  number of pit openings   94
  resistance to diffusion   161, 162
  resistance to flow   85, 90, 91
  structure   36–40
Transport processes (see Diffusion; Diffusion coefficient; Permeability; Thermal conductivity; Unsteady-state equations; Unsteady-state transport)
Tsuga canadensis
  bordered pit, TEM   49
  earlywood and latewood tracheids, TEM   38
  solvent-exchange drying   116
Turbulent flow
  equations   83
  relationship to energy consumption   83

relationship to radius 83
velocity profile 83
Tyloses 57, 63

U-value 146–148
Unsteady-state equations (see also Diffusion; Diffusion coefficients; Unsteady-state transport)
  diffusion
    isothermal 178, 179
    nonisothermal 221–223
  heat transfer 175–178, 223, 224
  gas flow
    cylinders 183
    parallel-sided bodies 179–183
      Petty model 182
      Sebastian equation 182
  liquid flow 213–220
    Bramhall model 216, 217
    cylinders 215, 218, 219
    parallel-sided bodies 214–217
    rectangular sections 219
  graphical solutions with constant coefficients
    E vs. τ, cylinders 191, 192
    E vs. τ, parallel-sided bodies 187, 188
    $\bar{E}$ vs. τ, cylinders 193
    $\bar{E}$ vs. τ, parallel-sided bodies 189
    F_{VL} vs. τ 215
  heat transfer, lumber 205, 207
  heat transfer, poles 204
  heat transfer, timbers 208, 209
Unsteady-state transport (see also Unsteady-state equations)
  definition 175
  diffusion 178, 179
  directional significance of flow 195–202
  graphs
    E, cylinder 201
    E, plane-surfaced body 200
    $\bar{E}$ or F_{VL}, cylinder 198
    $\bar{E}$ or F_{VL}, parallel-sided body 197
    F_g, gas flow, cylinder 202
  retention of liquids 210–213

Vapometer 171–174
Vapor barriers 24
  evaluation 173, 174
  suitability of wood 174
Vascular tracheids 54
Vasicentric tracheids 55, 56, 60–63
Vessels 57
  concentration of 57
  dimensions 53–56
  flow path 54, 61–63
  length 57, 58
  occlusion 57, 63
Viscosity
  air and water 78
  definition 77–79
Viscous flow (see also Darcy's law; Poiseuille's law; Permeability; Reynolds' number) 73–82
  effect of area to volume ratio 74, 77, 82
  relationship to energy consumption 83
  relationship to radius 81, 82
  velocity profile 80

Warty layer 37–39
Water potential (see also Chemical potential; Equilibrium moisture content; Fiber saturation point; Sorption isotherm)
  components 118, 119
  equations 119–122
  $h > 0.95$ 121, 122
  graphs
    soil 129
    wood 127
    unsaturated air 120
  potential for rise of sap in trees 128, 131
  relationship to chemical potential 119
  relationship to EMC and FSP 124–128
  temperature dependency 119
  values corresponding to FSP 125, 126
Wetting and nonwetting liquids 106–109
Wetting angle 107–109
Wick action 128
Wood properties (see also Diffusion; Heat transfer; Permeability; Unsteady-state transport)
  condensation inside solid wood 174
  fire resistance 133
  insulation 133, 148, 223–225
  nonisothermal diffusion 165–171, 221–223
  thermal capacity 225
  vapor barrier 174

Xylan 66–68, 71, 72

M. H. Zimmermann

Xylem Structure and the Ascent of Sap

1983. 64 figures. X, 142 pages
ISBN 3-540-12268-0

Contents: Introduction. – Conducting Units: Tracheids and Vessels. – The Vessel Network in the Stem. – The Cohesion Theory of Sap Ascent. – Hydraulic Architecture of Plants. – Other Functional Adaptations. – Failure and "Senescence" of Xylem Function. – Pathology of the Xylem.

The result of many years of interdisciplinary studies, this book is the first to provide an integrated description of sap ascension from an anatomical and functional point of view. The book opens with a consideration of the threedimensional aspects of wood anatomy. The cohesion theory is then introduced and a number of functional parameters are discussed, such as the tensible strength of water, water storage, sealing concepts, pressure gradients, and velocities. This is followed by a description of the overall xylem structure of plants – their so-called "hydraulic architecture." Other functional xylem adaptations are also covered, and the text concludes with a description of xylem and pathology.
Although the book contains over 400 citations, it is not intended to be encyclopedic in nature: it is an idea book that looks at old problems in new ways and highlights fascinating areas of current research.

Springer-Verlag
Berlin
Heidelberg
New York
Tokyo

W. H. Smith

Air Pollution and Forests

Interactions between Air Contaminants and Forest Ecosystems

1981. 60 figures. XV, 379 pages
(Springer Series on Environmental Management)
ISBN 3-540-90501-4

Contents: Introduction. – Forests Function as Sources and Sinks for Air Contaminants – Class I Interactions. – Forests Are Influenced by Air Contaminants in a Subtile Manner – Class II Interactions. – Forest Ecosystems Are Influenced by Air Contaminants in a Dramatic Manner – Class III Interactions. – Appendix: Common and Scientific Names of Woody Plants Cited in this Book. – Index.

Air Pollution and Forests is a comprehensive synthesis and overview of the complex interactions between air pollution and forest ecosystems. The author provides an inventory of all the significant relationships between forests and pollution, reviewing the sources of and sinks for air contaminants. Discussions center on the role of forests in major elementscycles and as sources of hydrocarbons and particulates. The influence of pollutants on tree reproduction, symbiotic microorganisms, and on photosynthesis is examined. The effects of forest fires and nutrient cycling throughout the system are also explored. Fine Finally, a topic of great economic and environmental concern, forest destruction, is presented from a biochemical as well as ecological viewpoint.

Springer-Verlag
Berlin
Heidelberg
New York
Tokyo